Linear Equation

A complete workbook with lessons and problems

By Maria Miller

Contents

Preface

Hello! I am Maria Miller, the author of this math book. I love math, and I also love teaching. I hope that I can help you to love math also!

I was born in Finland, where I also grew up and received all of my education, including a Master's degree in mathematics. After I left Finland, I started tutoring some home-schooled children in mathematics. That was what sparked me to start writing math books in 2002, and I have kept on going ever since.

In my spare time, I enjoy swimming, bicycling, playing the piano, reading, and helping out with Inspire4.com website. You can learn more about me and about my other books at the website MathMammoth.com.

This book, along with all of my books, focuses on the conceptual side of math... also called the "why" of math. It is a part of a series of workbooks that covers all math concepts and topics for grades 1-7. Each book contains both instruction and exercises, so is actually better termed *worktext* (a textbook and workbook combined).

My lower level books (approximately grades 1-5) explain a lot of mental math strategies, which help build number sense — proven in studies to predict a student's further success in algebra.

All of the books employ visual models and exercises based on visual models, which, again, help you comprehend the "why" of math. The "how" of math, or procedures and algorithms, are not forgotten either. In these books, you will find plenty of varying exercises which will help you look at the ideas of math from several different angles.

I hope you will enjoy learning math with me!

Introduction

Linear Equations Workbook presents the student with the basics of solving linear equations, including equations that involve a variable on both sides and equations that require the usage of the distributive property to eliminate parentheses. We also briefly study inequalities and graphing. This workbook best suits pre-algebra or grades 7 to 8 mathematics studies.

The first lesson reviews the concept of an equation and how to model equations using a pan balance (scale). The basic principle for solving equations is that, when you perform the same operation on both sides of an equation, the two sides remain equal.

The workbook presents two alternatives for keeping track of the operations to be performed on an equation. The one method, writing the operation under each side of the equation, is common in the United States. The other method, writing the operation in the right margin, is common in Finland. Either way is correct, and the choice is just a matter of the personal preference of the teacher.

The introduction to solving equations is followed by a lesson on addition and subtraction equations and another on multiplication and division equations. All the equations are easily solved in only one step of calculations. The twofold goal is to make the student proficient in manipulating negative integers and also to lay a foundation for handling more involved equations that are studied later on in the workbook.

In the next lesson, students write equations to solve simple word problems. Even though they could solve most of these problems without using the equations, the purpose of the lesson is to make the student proficient in writing simple equations before moving on to more complex equations from more difficult word problems.

The next topic, in the lesson *Constant Speed*, is solving problems with distance (d), rate or velocity (v), and time (t). Students use the equivalent formulas $d = vt$ and $v = d/t$ to solve problems involving constant or average speed. They learn an easy way to remember the formula $v = d/t$ from the unit for speed that they already know, "miles per hour."

In later lessons, we delve deeper into our study of equations. Now the equations require two or more steps to solve and may contain parentheses. The variable may appear on both sides of the equation. Students will also write equations to solve simple word problems.

There is also a lesson on patterns of growth, which may seem to be simply a fascinating topic, but in reality presents the fundamentals of a very important concept in algebra — that of linear functions (although they are not mentioned by that name)—and complements the study of lines in the subsequent lessons.

After the section about equations, the text briefly presents the basics of inequalities and how to graph them on a number line. Students apply the principles for solving equations to solve simple inequalities and word problems that involve inequalities.

The last major topic is graphing. Students begin the section by learning to graph linear equations and continue on to the concept of slope, which in informal terms is a measure of the inclination of a line. More formally, slope can be defined as the ratio of the change in y-values to the change in x-values. The final lesson applies graphing to the previously-studied concepts of speed, time, and distance through graphs of the equation $d = vt$ in the coordinate plane.

I wish you success in teaching math!

Maria Miller, the author

Helpful Resources on the Internet

Simplifying Expressions

Factor the Expressions Quiz
Factor expressions. For example, $-4x + 16$ factors into $-4(x - 4)$.
http://www.thatquiz.org/tq-0/?-jh00-l4-p0

Simplifying Algebraic Expressions Practice Problems
Practice simplifying expressions such as $4(2p - 1) - (p + 5)$ with these 10 questions. Answer key included.
http://www.algebra-class.com/algebraic-expressions.html

Distributive Property with Negative Numbers
Use the distributive property to remove the parentheses in this interactive exercise. Click to see an example.
http://www.hstutorials.net/dialup/distributiveProp.htm

Simplifying Algebraic Expressions (1)
Eight practice problems that you can check yourself about combining like terms and using the distributive property.
http://www.algebralab.org/lessons/lesson.aspx?file=Algebra_BasicOpsSimplifying.xml

Simplifying Algebraic Expressions (3)
You can check this five-question quiz from Glencoe yourself.
http://www.glencoe.com/sec/math/studytools/cgi-bin/msgQuiz.php4?isbn=0-07-825200-8&chapter=3&lesson=2&&headerFile=4

Equations

The Simplest Equations - Video Lessons by Maria
A set of free videos that teach the topics in this workbook - by the author.
http://www.mathmammoth.com/videos/prealgebra/pre-algebra-videos.php#equations

Stable Scales Quiz
In each picture, the scales are balanced. Can you find the weight of the items on the scales?
http://www.transum.org/software/SW/Starter_of_the_day/Students/Stable_Scales_Quiz.asp

Model Algebra Equations
Model an equation on a balance using algebra tiles (tiles with numbers or the unknown x). Then, solve the equation by placing -1 tiles on top of $+1$ tiles or vice versa. Includes one-step and two-step equations.
http://www.mathplayground.com/AlgebraEquations.html

Balance When Adding and Subtracting
Click on the buttons above the scales to add or subtract until you can figure out the value of x in the equation.
http://www.mathsisfun.com/algebra/add-subtract-balance.html

One-Step Equations Quizzes
Practice one-step equations in these timed quizzes.
http://crctlessons.com/One-Step-Equations/one-step-equations.html

http://crctlessons.com/One-Step-Equation-Test/one-step-equation-test.html

Modeling with One-Step Equations
Practice writing basic equations to model real-world situations in this interactive activity from Khan Academy.
https://www.khanacademy.org/math/pre-algebra/pre-algebra-equations-expressions/pre-algebra-equation-word-problems/e/equations-in-one-variable-1

Exploring Equations E-Lab
Choose which operation to do to both sides of an equation in order to solve one-step equations.
http://www.harcourtschool.com/activity/elab2004/gr6/12.html

Algebra Four
A connect four game with equations. For this level, choose difficulty "Level 1" and "One-Step Problems".
http://www.shodor.org/interactivate/activities/AlgebraFour/

Algebra Meltdown
Solve simple equations using function machines to guide atoms through the reactor. Don't keep the scientists waiting too long or they blow their tops.
http://www.mangahigh.com/en/games/algebrameltdown

One-Step Equation Game
Choose the correct root for the given equation (multiple-choice), and then you get to attempt to shoot a basket.
http://www.math-play.com/One-Step-Equation-Game.html

Arithmagons
Find the numbers that are represented by question marks in this interactive puzzle.
http://www.transum.org/Software/SW/Starter_of_the_day/starter_August20.ASP

Cars
Use clues to help you find the total cost of four cars in this fun brainteaser.
http://www.transum.org/Software/SW/Starter_of_the_day/starter_July16.ASP

Two-Step Equations

Model Algebraic Equations with a Scale
Model and solve algebraic equations using a pan balance and tiles. Choose "2-Step Equations" for this level.
http://www.mathplayground.com/AlgebraEquations.html

Two-Step Equations Game
Choose the correct root for the given equation (multiple-choice), and then you get to attempt to shoot a basket. The game can be played alone or with another student.
http://www.math-play.com/Two-Step-Equations-Game.html

Solving Two-Step Equations
Type the answer to two-step-equations such as $-4y + 9 = 29$, and the computer checks it. If you choose "Practice Mode," it is not timed.
http://www.xpmath.com/forums/arcade.php?do=play&gameid=64

Two-Step Equations
Practice solving equations that take two steps to solve in this interactive exercise from Khan Academy.
https://www.khanacademy.org/math/algebra/one-variable-linear-equations/alg1-two-steps-equations-intro/e/linear_equations_2

Two-Step Equations Word Problems
Practice writing equations to model and solve real-world situations in this interactive exercise.
https://www.khanacademy.org/math/algebra-basics/alg-basics-linear-equations-and-inequalities/alg-basics-two-steps-equations-intro/e/linear-equation-world-problems-2

Visual Patterns
Hundreds of growing patterns. The site provides the answer to how many elements are in step 43 of the pattern.
http://www.visualpatterns.org/

More Equations

Solve Equations Quiz
A 10-question online quiz where you need to solve equations with an unknown on both sides.
http://www.thatquiz.org/tq-0/?-j102-l4-p0

Equations Level 3 Online Exercise
Practice solving equations with an unknown on both sides in this self-check online exercise.
http://www.transum.org/software/SW/Starter_of_the_day/Students/Equations.asp?Level=3

Missing Lengths
Try to figure out the value of the letters used to represent the missing numbers.
http://www.transum.org/software/SW/Starter_of_the_day/Students/Missing_Lengths.asp

Equations Level 4 Online Exercise
Practice solving equations which include brackets in this self-check online exercise.
http://www.transum.org/software/SW/Starter_of_the_day/Students/Equations.asp?Level=4

Equations Level 5 Online Exercise
This exercise includes more complex equations requiring multiple steps to find the solution.
http://www.transum.org/software/SW/Starter_of_the_day/Students/Equations.asp?Level=5

Solving Equations Quizzes
Here are some short online quizzes that you can check yourself.
http://www.glencoe.com/sec/math/studytools/cgi-bin/msgQuiz.php4?isbn=0-07-825200-8&chapter=3&lesson=5&&headerFile=4

http://www.phschool.com/webcodes10/index.cfm?wcprefix=bja&wcsuffix=0701

Rags to Riches Equations
Choose the correct root to a linear equation.
http://www.quia.com/rr/4096.html

Solve Equations Exercises
Click "new problem" (down the page) to get a randomly generated equation to solve. This exercise includes an optional graph which the student can use as a visual aid.
http://www.onemathematicalcat.org/algebra_book/online_problems/solve_lin_int.htm#exercises

Equation Word Problems Quiz
Solve word problems which involve equations and inequalities in this multiple-choice online quiz.
http://www.phschool.com/webcodes10/index.cfm?wcprefix=bja&wcsuffix=0704

Whimsical Windows - Equation Game
Write an equation for the relationship between x and y based on a table of x and y values. Will you discover the long lost black unicorn stallion?
http://mrnussbaum.com/whimsical-windows/

Inequalities

Inequality Quiz
A 10-question multiple choice quiz on linear inequalities (like the ones studied in this workbook).
http://www.mrmaisonet.com/index.php?/Inequality-Quiz/Inequality-Quiz.html

Inequalities
Here is another five-question quiz from Glencoe that you can check yourself.
http://www.glencoe.com/sec/math/studytools/cgi-bin/msgQuiz.php4?isbn=0-07-825200-8&chapter=7&lesson=3&&headerFile=4

Plot Simple Inequalities
Practice plotting simple inequalities on a number line in this 10-question interactive quiz.
https://www.thatquiz.org/tq-o/?-j18-l1-p0

Match Inequalities and Their Plots
Match the statements with the corresponding diagrams in this interactive online activity.
http://www.transum.org/software/SW/Starter_of_the_day/Students/InequalitiesB.asp?Level=5

Solve Simple Inequalities
For each inequality, find the range of values for x which makes the statement true. An example is given.
http://www.transum.org/software/SW/Starter_of_the_day/Students/InequalitiesC.asp?Level=6

Two-Step Inequality Word Problems
Practice constructing, interpreting, and solving linear inequalities that model real-world situations.
https://www.khanacademy.org/math/algebra/one-variable-linear-inequalities/alg1-two-step-inequalities/e/interpretting-solving-linear-inequalities

Speed, Time, and Distance

Distance, Speed, and Time from BBC Bitesize
Instruction, worked out exercises, and an interactive quiz relating to constant speed, time, and distance.
A triangle with letters D, S, and T helps students remember the formulas for distance, speed, and time.
http://www.bbc.co.uk/bitesize/standard/maths_i/numbers/dst/revision/1/

Absorb Advanced Physics - Speed
An online tutorial that teaches the concept of average speed with the help of interactive simulations and exercises.
http://www.absorblearning.com/advancedphysics/demo/units/010101.html#Describingmotion

Understanding Distance, Speed, and Time
An interactive simulation of two runners. You set their starting points and speeds, and observe their positions as the tool runs the simulation. It graphs the position of both runners in relation to time.
http://illuminations.nctm.org/Activity.aspx?id=6378

Representing Motion
A tutorial an interactive quiz with various questions about speed, time, and distance.
http://www.bbc.co.uk/schools/gcsebitesize/science/add_aqa_pre_2011/forces/represmotionrev1.shtr

Distance-Time Graphs
An illustrated tutorial about distance-time graphs. Multiple-choice questions are included.
http://www.absorblearning.com/advancedphysics/demo/units/010103.html

Distance-Time Graph
Click the play button to see a distance-time graph for a vehicle which moves, stops, and then changes direction.
http://www.bbc.co.uk/schools/gcsebitesize/science/add_aqa_pre_2011/forces/represmotionrev5.shtml

Distance versus Time Graph Puzzles
Try to move the stick man along a number line in such a way as to illustrate the graph that is shown.
http://davidwees.com/graphgame/

Graphing and Slope

Graph Linear Equations
A ten-question online quiz where you click on three points on the coordinate grid to graph the given equation.
http://www.thatquiz.org/tq-0/?-j10g-l4-p0

Find the Slope
A ten-question online quiz that asks for the slope of the given line.
http://www.thatquiz.org/tq-0/?-j300-l4-p0

Slope Slider
Use the sliders to change the slope and the *y*-intercept of a linear equation to see what effect they have on the graph of the line.
http://www.shodor.org/interactivate/activities/SlopeSlider/

Graphing Equations Match
Match the given equations to their corresponding graphs.
http://www.math.com/school/subject2/practice/S2U4L3/S2U4L3Pract.html

Find Slope from Graph
Find the slope of a line on the coordinate plane in this interactive online activity.
https://www.khanacademy.org/math/algebra-basics/alg-basics-graphing-lines-and-slope/alg-basics-slope/e/slope-from-a-graph

Slope - Exercises
Practice finding the slope in this interactive online exercise.
http://www.onemathematicalcat.org/algebra_book/online_problems/compute_slope.htm#exercises

Graphs Quiz
Check your knowledge of graphing with this interactive self-check quiz.
http://www.glencoe.com/sec/math/studytools/cgi-bin/msgQuiz.php4?isbn=0-07-860467-2&chapter=5&lesson=2&headerFile=4

Equations and Graphing Quiz
Practice linear equations and functions in this interactive online test.
http://www.glencoe.com/sec/math/studytools/cgi-bin/msgQuiz.php4?isbn=0-07-829633-1&chapter=4&headerFile=4

General

Algebra Quizzes
A variety of online algebra quizzes from MrMaisonet.com.
http://www.mrmaisonet.com/index.php?/Algebra-Quizzes/

Pre-algebra Quizzes
Pearson provides a variety of online algebra quizzes to support their *Algebra Readiness* textbook.
http://www.phschool.com/webcodes10/index.cfm?fuseaction=home.gotoWebCode&wcprefix=bjk&wcsuffix=0099

Solving Equations

Do you remember? An **equation** has two expressions, separated by an equal sign:

(expression) = (expression)

To solve an equation, we can

- ¿ add the same quantity to both sides
- ¿ subtract the same quantity from both sides
- ¿ multiply both sides by the same number
- ¿ divide both sides by the same number

Notice that in any of these operations, the two expressions on the left and right sides of the equation will remain equal, even though the expressions themselves change!

Example. We will manipulate the simple equation $2 + 3 = 5$ in these four ways. We will write in the margin the operation that is going to be done next to both sides.

Let's add six to both sides. $\qquad\qquad\qquad\qquad$ $2 + 3 = 5$ $\qquad$ $| + 6$

Now, both sides equal 11. Next, we multiply both sides by 8. $\qquad$ $2 + 3 + 6 = 11$ $\qquad$ $\cdot\, \mathbf{8}$

Now, both sides equal 88. Next, we subtract 12 from both sides. $\qquad$ $16 + 24 + 48 = 88$ $\qquad$ $-\, \mathbf{12}$

Now both sides equal 76. Next, we divide both sides by 2. $\qquad$ $16 + 24 + 48 - 12 = 76$ $\qquad$ $\div\, \mathbf{2}$

Now both sides equal 38. $\qquad\qquad\qquad\qquad$ $8 + 12 + 24 - 6 = 38$

Of course, you do not usually work with equations like the one above, but with ones that have an unknown. Your goal is to **isolate** the unknown, or **leave it by itself,** on one side. Then the equation is solved.

We can model an equation with a **pan balance**. Both sides (pans) of the balance will have an *equal* weight in them, thus the sides are balanced (not tipped to either side).

Example. Solve the equation $x - 2 = 3$.

We can write this equation as $x + (-2) = 3$ and model it using negative and positive counters in the balance.

Here x is accompanied by two negatives on the left side. Adding two <u>positives</u> *to both sides* will cancel those two negatives. We denote that by writing "+2" in the margin.

margin ↓

$x + (-2) = 3$ $\qquad$ $| + 2$

We write $x + (-2) + 2 = 3 + 2$ to show that 2 was added to both sides of the equation.

$x + (-2) + 2 = 3 + 2$

Now the two positives and two negatives on the left side cancel each other, and x is left by itself. On the right side we have 5, so x equals 5 positives.

$x = 5$

1. Solve the equations. Write in the margin what operation you do to both sides.

a. Balance	Equation	Operation to do to both sides
	$x + 1 = -4$	

b. Balance	Equation	Operation to do to both sides
	$x - 1 = -3$	

c. Balance	Equation	Operation
	$x - 2 = 6$	

d. Balance	Equation	Operation
	$x + 5 = 2$	

2. If you need more practice, solve the following equations also. Draw a balance in your notebook to help you.

a. $x + (-3) = 7$ **b.** $x + (-3) = -4$ **c.** $x + 6 = -1$ **d.** $x + 5 = -4$

In the two examples below, we either multiply or divide both sides by the same number. Study them carefully!

Example 1.	Example 2.

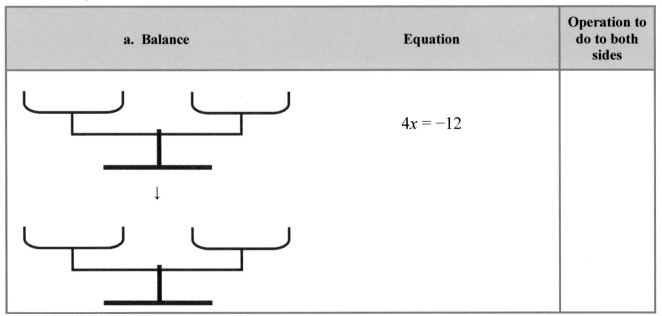

$3x = -9 \quad | \div 3$

$x = -3$

$(1/2)x = -3 \quad | \cdot 2$

$x = -6$

3. Solve the equations. Write in the margin what operation you do to both sides.

a. Balance	Equation	Operation to do to both sides
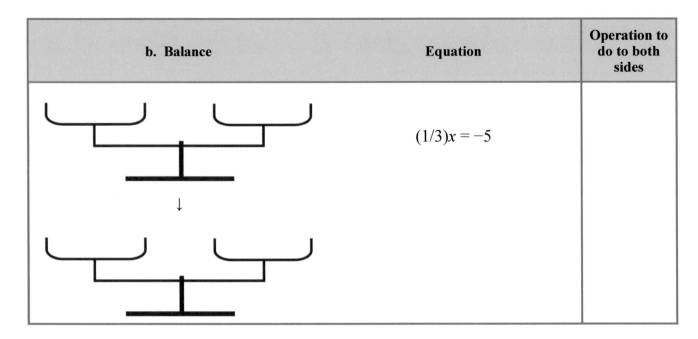	$4x = -12$	

b. Balance	Equation	Operation to do to both sides
	$(1/3)x = -5$	

16

4. Let's review a little! Which equation matches the situation?

a. A stuffed lion costs $8 less than a stuffed elephant.
Note: p_l signifies the price of the lion and p_e the price of the elephant.

$p_e = 8 - p_l$	$p_e = p_l - 8$	$p_l = p_e - 8$

b. A shirt is discounted by 1/5, and now it costs $16.

$p - 1/5 = \$16$	$\dfrac{4p}{5} = \$16$	$\dfrac{p}{5} = \$16$	$\dfrac{5p}{4} = \$16$	$\dfrac{p}{5} = 4 \cdot \$16$

5. Find the roots of the equation $\dfrac{6}{x+1} = -3$ in the set $\{2, -2, 3, -3, 4, -4\}$.

6. Write an equation, then solve it using guess and check. Each root is between -20 and 20.

a. 7 less than x equals 5.
b. 5 minus 8 equals x plus 1.
c. The quantity x minus 1 divided by 2 is equal to 4.
d. x cubed equals 8.
e. -3 is equal to the quotient of 15 and y.
f. Five times the quantity x plus 1 equals 10.

Example. Solve $2x + 5 = -3$.
The solution requires two steps.

The two x's are accompanied by five positives. Therefore, we will subtract five *from both sides*.

Subtracting 5 is the same as adding -5, so the right side ends up with 8 negatives.

The positives and negatives on the left side cancel each other, and $2x$ is left by itself on that side.

Now we need to divide both sides by 2. Again, we note that in the margin.

We can see that x equals 4 negatives.

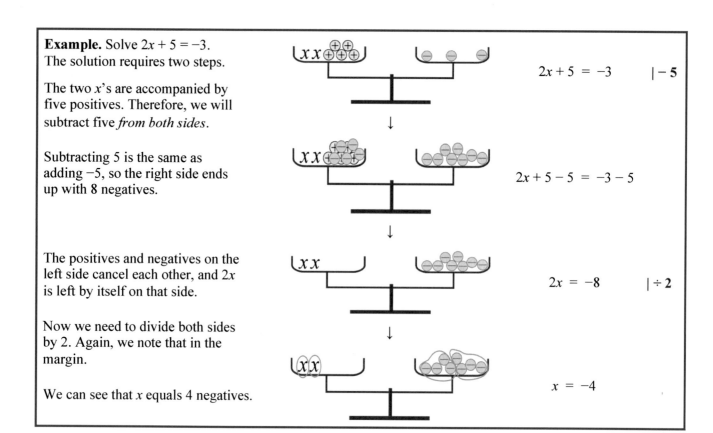

$2x + 5 = -3 \qquad | -5$

$2x + 5 - 5 = -3 - 5$

$2x = -8 \qquad | \div 2$

$x = -4$

7. Solve the equations. Write in the margin what operation you do to both sides.

a. Balance	Equation	Operation to do to both sides
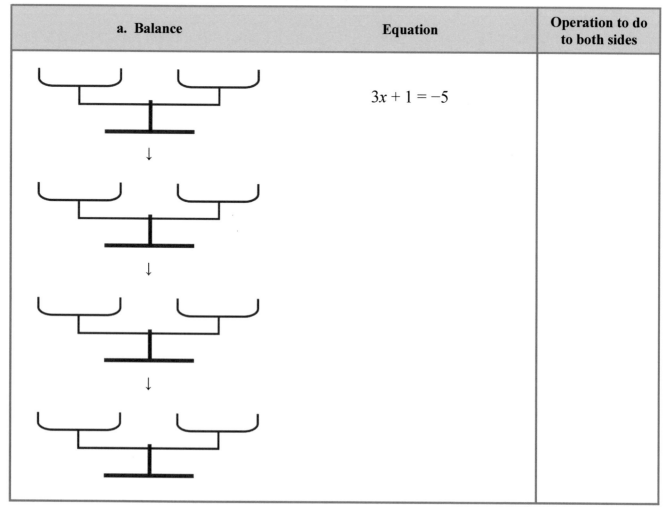	$3x + 1 = -5$	

18

b. Balance	Equation	Operation
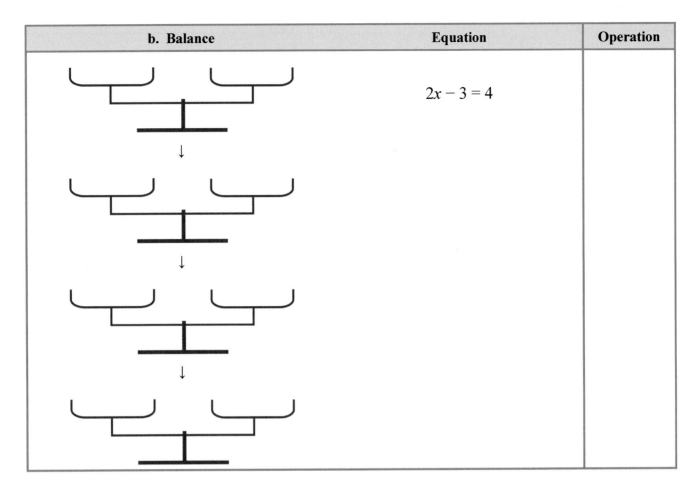	$2x - 3 = 4$	

c. Balance	Equation	Operation
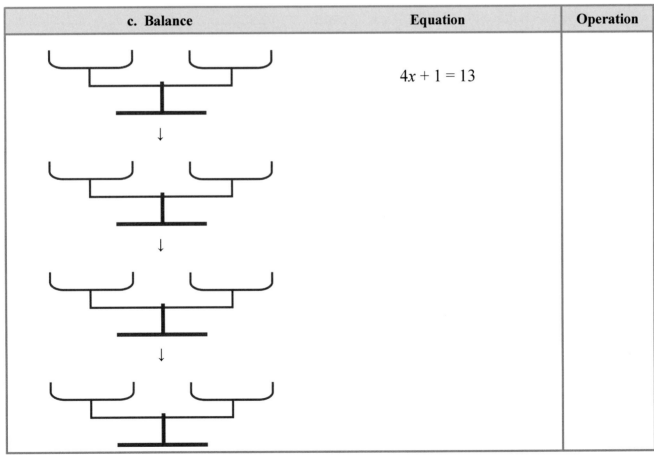	$4x + 1 = 13$	

Addition and Subtraction Equations

You can **keep track of the operations you're using in a couple of different ways.** One way is to write the operation underneath the equation on both sides. Another is to write it in the right margin, like we did in the last lesson.	**One way:** $x + 9 = 4$ $\underline{-9 \quad\quad -9}$ $x = -5$	**Another way:** $x + 9 = 4$　$\mid -9$ $x + 9 - 9 = 4 - 9$　(This step is optional.) $x = -5$
But in either case, always check your solution: does it solve the original equation?	Check: $-5 + 9 \overset{?}{=} 4$. Yes, it checks.	

1. Solve these one-step equations. Keep track of the operations either under the equation or in the margin, whichever way your teacher prefers.

a.	$x + 5 = 9$	**b.**	$x + 5 = -9$
c.	$x - 2 = 3$	**d.**	$w - 2 = -3$
e.	$z + 5 = 0$	**f.**	$y - 8 = -7$

2. In these equations, your first step is to <u>simplify</u> what is on the right side.

a.	$x - 7 = 2 + 8$ $x - 7 = 10$	**b.**	$x - 10 = -9 + 5$
c.	$s + 5 = 3 + (-9)$	**d.**	$t + 6 = -3 - 5$

If the **unknown is on the right side of the equation**, you have two options:	**1. Flip the sides first:**	**2. Solve as it is:**
¿ First, flip the two sides. Then solve as usual. ¿ Or, solve as usual, isolating the unknown —this time on the right side of the equation. The solution will initially read as $-7 = x$. Flip the sides now and write the solution as $x = -7$.	$-9 = x - 2$ $x - 2 = -9$ <u>$+2 \qquad +2$</u> $x = -7$	$-9 = x - 2$ <u>$+2 \qquad +2$</u> $-7 = x$ $x = -7$

3. Solve. Check your solutions.

a. $\qquad -8 = s + 6$	**b.** $\qquad -2 = x - 7$
c. $\qquad 4 = s + (-5)$	**d.** $\qquad 2 - 8 = y + 6$
e. $\qquad 5 + x = -9$	**f.** $\qquad -6 - 5 = 1 + z$
g. $\qquad y - (-7) = 1 - (-5)$	**h.** $\qquad 6 + (-2) = x - 2$
i. $\qquad 3 - (-9) = x + 5$	**j.** $\qquad 2 - 8 = 2 + w$

Example 1. Solve $-2 + 8 = -x$. Our first step is to simplify the sum $-2 + 8$. The equation becomes $-x = 6$ (or $6 = -x$). What does that mean? It means that the <u>opposite of x</u> is 6. So x must equal -6! Lastly we check the solution $x = -6$: $-2 + 8 \overset{?}{=} -(-6)$, which simplifies to $6 \overset{?}{=} 6$, so it checks.	**1. Flip the sides first:** $-2 + 8\ =\ -x$ $-x\ =\ -2 + 8$ $-x\ =\ 6$ $x\ =\ -6$	**2. Solve as it is:** $-2 + 8\ =\ -x$ $6\ =\ -x$ $-6\ =\ x$ $x\ =\ -6$

4. Solve for x. Check your solutions.

a. $\qquad -x\ =\ 6$	**b.** $\qquad -x\ =\ 5 - 9$
c. $\qquad 4 + 3\ =\ -y$	**d.** $\qquad -2 - 6\ =\ -z$

5. Which equation best matches the situation?

   a. The sides of a square playground were shortened by 1/2 m, and now its perimeter is 12 m.

$4s - 1/2 = 12$	$4(s - 1/2) = 12$	$4s - 50 = 12$
$s - 1/2 = 4 \cdot 12$	$s - 1/2 = 12$	$4(s - 0.5) = 12$

   b. How long were the sides before they were made shorter? Solve the problem using mental math.

   c. *Challenge*: Solve the same problem using the equation. Compare the steps of this formal solution to the way you reasoned it out in your head. Are the steps similar?

6. Here is another "growing pattern." Draw steps 4 and 5 and answer the questions.

Step   1   2   3

a. How do you see this pattern grow?

b. How many flowers will there be in step 39?

c. In step n?

Example 2. Solve $8 - x = -2$.

As usual, think about what you need to do to isolate the x on one side. Since there is an 8 on the side with the x, we need to subtract 8 from both sides.

However, note that x is being *subtracted*, or in other words, there is a negative sign in front of the x.

This negative sign <u>does not disappear</u> when you subtract 8 from both sides.

Writing the operation underneath each side:	Writing the operation in the right margin:
$$\begin{aligned} 8 - x &= -2 \\ \underline{-8} \quad &\underline{-8} \\ -x &= -10 \\ x &= 10 \end{aligned}$$	$$\begin{aligned} 8 - x &= -2 \quad \big|{-8} \\ 8 - x - 8 &= -2 - 8 \\ -x &= -10 \\ x &= 10 \end{aligned}$$

If this is confusing, think of it this way: The equation $8 - x = -2$ can also be written as $8 + (-x) = -2$. When we subtract 8 from both sides, the left side becomes $8 + (-x) - 8$. The positive 8 and negative 8 will cancel each other and leave $-x$.

So we end up with the equation $-x = -10$. This equation says that the opposite of x is negative 10, so x must be 10. (Why?)

Lastly, check your solution by substituting $x = 10$ back into the original equation: $8 - \underline{10} \overset{?}{=} -2$ ✓

7. Solve. Check your solutions.

a. $\quad 2 - x = 6$	**b.** $\quad 8 - x = 7$
c. $\quad -5 - x = 5$	**d.** $\quad 2 - x = -6$
e. $\quad 1 = -5 - x$	**f.** $\quad 2 + (-9) = 8 - z$
g. $\quad -8 + r = -5 + (-7)$	**h.** $\quad 2 - (-5) = 2 + 5 + t$

Multiplication and Division Equations

Do you remember **how to show simplification**? Just cross out the numbers and write the new numerator above the fraction and the new denominator below it. Notice that the number you divide by (the 5 in the fraction at the right) isn't indicated in any way!	$\dfrac{\overset{7}{\cancel{35}}}{\underset{11}{\cancel{55}}} = \dfrac{7}{11}$
We can simplify expressions involving variables in exactly the same way. In the examples on the right, we cross out the *same number* from the numerator and the denominator. That is based on the fact that a number divided by itself is 1. We could write a little "1" beside each number that is crossed out, but that is usually omitted.	$\dfrac{\cancel{2}x}{\cancel{2}} = x \qquad \dfrac{\cancel{5}s}{\cancel{5}} = s$ $\dfrac{4\cancel{x}}{\cancel{x}} = 4$
In this example, we simplify the fraction 3/6 into 1/2 the usual way.	$\dfrac{\overset{1}{\cancel{3}}x}{\underset{2}{\cancel{6}}} = \dfrac{1}{2}x \ \text{ or } \ \dfrac{x}{2}$
Notice: We divide both the numerator and the denominator by 8, but <u>this leaves −1 in the denominator</u>. Therefore, the whole expression simplifies to −z instead of z.	$\dfrac{\cancel{8}z}{\cancel{-8}} = \dfrac{z}{-1} = -z$

1. Simplify.

a. $\dfrac{8x}{8}$	**b.** $\dfrac{8x}{2}$	**c.** $\dfrac{2x}{8}$
d. $\dfrac{-6x}{-6}$	**e.** $\dfrac{-6x}{6}$	**f.** $\dfrac{6x}{-6}$
g. $\dfrac{6w}{2}$	**h.** $\dfrac{6w}{w}$	**i.** $\dfrac{6w}{-2}$

2. Draw the fourth and fifth steps of the pattern and answer the questions.

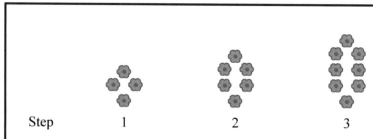

Step   1   2   3

a. How would you describe the growth of this pattern?

b. How many flowers will there be in step 39?

c. In step *n*?

Now you should be ready to use multiplication and division to solve simple equations.

Example 1. Solve $-2x = 68$.	$-2x = 68$ This is the original equation.

Example 1. Solve $-2x = 68$.

The unknown is being multiplied by -2. To isolate it, we need to divide both sides by -2. (See the solution on the right.)

We get $x = -34$. Lastly we check the solution by substituting -34 in the place of x in the original equation:

$$-2(-34) \stackrel{?}{=} 68$$

$$68 = 68 \quad \text{It checks.}$$

$-2x = 68$ This is the original equation.

$\dfrac{-2x}{-2} = \dfrac{68}{-2}$ We divide both sides by -2.

$\dfrac{\cancel{-2}x}{\cancel{-2}} = \dfrac{68}{-2}$ Now it is time to simplify. We cross out the -2 factors on the left side. On the right side, we do the division.

$x = -34$ This is the final answer.

Note: Most people combine the first 3 steps into one when writing the solution. Here they are written out for clarity.

3. Solve. Check your solutions.

a. $5x = -45$	**b.** $-3y = -21$
c. $-4 = 4s$	**d.** $72 = -6y$

4. Solve. Simplify the one side first.

a. $-5q = -40 - 5$	**b.** $2 \cdot 36 = -6y$
c. $3x = -4 + 3 + (-2)$	**d.** $5 \cdot (-4) = -10z$

Example 2. Solve $\dfrac{x}{-6} = -5$.

Here the unknown is divided by −6. To undo that division, we need to *multiply* both sides by −6. (See the solution on the right.)

We get $x = 30$. Lastly we check the solution:

$$\dfrac{30}{-6} \overset{?}{=} -5$$

$$-5 = -5 \quad \checkmark$$

$\dfrac{x}{-6} = -5$	This is the original equation.
$\dfrac{x}{-6} \cdot (-6) = -5 \cdot (-6)$	We multiply both sides by −6.
$\dfrac{x}{\cancel{-6}} \cdot (\cancel{-6}) = 30$	Now it is time to simplify. We cross out the −6 factors on the left side, and multiply on the right.
$x = 30$	This is the final answer.

When writing the solution, most people would combine steps 2 and 3. Here both are written out for clarity.

5. Solve. Check your solutions.

a. $\dfrac{x}{2} = -45$	**b.** $\dfrac{s}{-7} = -11$
c. $\dfrac{c}{-7} = 4$	**d.** $\dfrac{a}{-13} = -9 + (-11)$

6. Write an equation for each situation. Then solve it. Do not write the answer only, as the main purpose of this exercise is to practice writing equations.

   a. A submarine was located at a depth of 500 ft.
   There was a shark swimming at 1/6 of that depth.
   At what depth is the shark?

   b. Three towns divided highway repair costs equally.
   Each town ended up paying $21,200.
   How much did the repairs cost in total?

Example 3. Solve $-\dfrac{1}{5}x = 2$. Here the unknown is multiplied by a negative fraction, but do not panic!

You see, you can *also* write this equation as $\dfrac{x}{-5} = 2$, where the unknown is simply divided by negative 5.

So what should we do in order to isolate x?

That is correct! Multiplying by -5 will isolate x. In the boxes below, this equation is solved in two slightly different ways, though both are doing essentially the same thing: multiplying both sides by -5.

Multiplying a fraction by its reciprocal:	**Canceling a common factor:**
$-\dfrac{1}{5}x \quad = \quad 2 \qquad \mid \cdot (-5)$	$-\dfrac{1}{5}x \quad = \quad 2 \qquad$ rewrite the equation
$(-5)\cdot\left(-\dfrac{1}{5}\right)x \;=\; (-5)\cdot 2 \quad$ Note that -5 times $-1/5$ is 1.	$\dfrac{x}{-5} \quad = \quad 2 \qquad \mid \cdot (-5)$
$1x \;=\; -10$	$\dfrac{x}{-5} \cdot (-5) \;=\; 2 \cdot (-5)$
$x \;=\; -10$	$x \;=\; -10$

Lastly we check the solution by substituting -10 in place of x in the original equation:

$$-\dfrac{1}{5}(-10) \overset{?}{=} 2$$
$$2 \;=\; 2 \qquad \text{It checks.}$$

7. Solve. Check your solutions.

a. $\quad \dfrac{1}{3}x \;=\; -15$	**b.** $\quad -\dfrac{1}{6}x \;=\; -20$	**c.** $\quad -\dfrac{1}{4}x \;=\; 18$
d. $\quad -2 \;=\; -\dfrac{1}{9}x$	**e.** $\quad -21 \;=\; \dfrac{1}{8}x$	**f.** $\quad \dfrac{1}{12}x \;=\; -7+5$

Word Problems 1

Example 1. The area of a rectangle is 195 m². One side measures 13 m. How long is the other side?

To write the equation, you need to remember that the area of any rectangle is calculated as **area = side · side** .

That relationship gives us our equation. We simply substitute the known area and the length of the known side into the equation and represent the length of the unknown side by some variable:

$$195 = s \cdot 13$$

Then we rewrite the expression $s \cdot 13$ in the usual way, where the coefficient (13) comes first and the variable(s) last. We also flip the sides of the equation, so the unknown is on the left: $13s = 195$.

Solution:

$$13s = 195$$

$$\frac{13s}{13} = \frac{195}{13} \qquad \text{We divide both sides by 13 and simplify.}$$

$$s = 15 \qquad \text{This is the solution.}$$

Check:

$$13 \cdot 15 \overset{?}{=} 195$$

$$195 = 195 \checkmark$$

For each given situation, write an equation and solve it. The problems themselves are simple, and you could solve them without writing an equation, but it is important to practice writing equations! You need to learn to write equations for simple situations now, so you will be able to write equations for more complex situations later on.

1. The perimeter of a square is 456 cm. How long is one side?

   Equation:

2. The area of a rectangular park is 4,588 square feet.
   One side measures 62 feet. How long is the other side?

   Equation:

3. John bought some boxes of screws for $15 each and paid a total of $165. How many boxes did he buy?

   Equation:

4. A candle burned at a steady rate of 2 cm per hour for 4 hours. Now it is only 6 cm long. How long was the candle at first?

   Equation:

5. A baby dolphin is only 1/12 as heavy as his mother. The baby weighs 15 kilograms. How much does the mother dolphin weigh?

   Equation:

Example 2. Solve the equation $45 + y + 82 + 192 = 374$ using a bar model and also using an equation.

Bar model:

For the bar model, we draw the addends as parts of the total, which we indicate with a double-headed arrow.

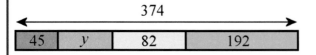

To solve it, we subtract all the known parts from 374:

$y = 374 - 45 - 82 - 192$

$y = 55$

Equation:

$45 + y + 82 + 192 = 374$

$319 + y = 374$

$319 + y - 319 = 374 - 319$

$y = 55$

First, add what you can on the left side.

Next, subtract 319 from both sides.

Check: $45 + 55 + 82 + 192 \overset{?}{=} 374$

$374 = 374$ ✔

6. Using both a bar model and an equation, solve the equation $21 + x + 193 = 432$.

Bar model:	Equation:

7. Using both a bar model and an equation, solve the equation $495 + 304 + w + 94 = 1,093$.

Bar model:	Equation:

Constant Speed

If an object travels with a constant speed, we have three quantities to consider: *speed or velocity (v), time (t),* and *distance (d).* The formula $d = vt$ tells us how they are interconnected.

Does that formula make sense?

Let's say John rides his bicycle at a constant speed of 12 km per hour for four hours. How far can he go?

The formula says you multiply the speed (12 km/h) by the time (4 h) to get the distance (48 km). So the formula does make sense — that is how our "common sense" tells us to calculate it also.

Example 1. A boat travels at a constant speed of 15 km/h. How long will it take the boat to go a distance of 21 km?

The problem gives us the speed and the distance. The time (*t*) is unknown.

We can solve the unknown time by using the formula $d = vt$. We simply substitute the given values of *v* and *d* in it, and we will get an equation that we can solve:

$$d = v\ t$$
$$\downarrow \qquad \downarrow$$
$$21 = 15\ t$$

To make it easier, we will leave off the units while solving the equation. We can do that, since both the velocity and the distance involve kilometers.

Next, we simply solve this equation:

21 =	15*t*	Flip the sides.
15*t* =	21	Divide both sides by 15.
$\dfrac{15t}{15}$ =	$\dfrac{21}{15}$	The 15 in numerator and denominator cancel.
t =	1 6/15	

The final answer *t* = 1 6/15 is in *hours*, because the unit for speed was kilometers per <u>hour</u>.

Let's work on that answer a bit more. First, it simplifies to 1 2/5 hours. Then let's change 2/5 of an hour to minutes.

How much is 1/5 of an hour? That's right, it is 12 minutes. And 2/5 of an hour is 24 minutes. So the final answer is 1 hour, 24 minutes.

1. Use the formula $d = vt$ to solve the problems.

a. A caterpillar crawls along at a constant speed of 20 cm/hour. How long will it take it to travel 34 cm?

$$d = v\ t$$
$$\downarrow \quad \downarrow \quad \downarrow$$

b. Father leaves at 7:40 a.m. to drive 20 miles to work. If his average speed is 48 mph, when will he arrive at work?

$$d = v\ t$$
$$\downarrow \quad \downarrow \quad \downarrow$$

How to change hours and minutes into fractional and decimal hours and vice versa	
Example 2. Change 14 minutes into hours. Since there are 60 minutes in an hour, 14 minutes is simply 14/60 of an hour. This simplifies to 7/30 hours. You can also change it into a decimal by dividing to get 0.233 hours (rounded to three decimals).	**Example 3.** Change 4.593 hours into hours and minutes. How many minutes are in the decimal part? Since one hour is 60 minutes, 0.593 hours is 0.593 · 60 minutes = 35.58 minutes ≈ 36 minutes. So 4.593 hours ≈ 4 hours 36 minutes.

2. Convert the given times into hours in decimal format. Round your answers to three decimal digits.

a. 35 minutes	**b.** 44 minutes
c. 2 h 16 min	**d.** 4 h 9 min

3. Give these times in hours and minutes.

a. 2.4 hours	**b.** 0.472 hours
c. 3 3/5 hours	**d.** 16/50 hours

4. The average speed of a bus is 64 km/hour. What distance can it travel in 4 hours and 15 minutes?

5. Sam is an athlete who can run 10 miles in an hour. How long will it take him to run home from the shopping center, a distance of 2.4 miles?

6. A train traveled a distance of 360 miles between two towns so that in the first half of the distance, its average speed was 90 mph, and in the second half, only 75 mph. How long did it take to travel from the one town to the other?

7. Elijah wants to use the extra time between classes for exercising. He plans to jog for 25 minutes in one direction, turn, and jog back to school. What is the distance Elijah can jog in 25 minutes if his average jogging speed is 6 mph?

Example 4. Andy drove from his home to his workplace, which was 22 miles away, in 26 minutes. What was his average speed?

The average speed of cars is usually given in miles per hour or kilometers per hour. This unit of speed actually gives us a **formula for calculating speed**:

$$\begin{array}{ccccc} \text{speed} & \text{is} & \text{miles} & \text{per} & \text{hour} \\ \downarrow & \downarrow & \downarrow & \downarrow & \downarrow \\ \text{velocity} & = & \text{distance} & / & \text{time} \end{array}$$

(The word "per" indicates division.)

In symbols, $v = \dfrac{d}{t}$.

His average speed is therefore

$$v \;=\; \frac{d}{t} \;=\; \frac{22 \text{ miles}}{26 \text{ minutes}} \;=\; \frac{11}{13} \text{ miles per minute}$$

$$\approx 0.846 \text{ miles per minute.}$$

The problem is that the average speed is usually given in miles per *hour*, not miles per minute. How can we fix that?

One way is to multiply our answer by 60. Doing that, we get 0.846 mi/min · 60 min/hr = 50.76 ≈ 51 mph.

Another way is to change the original time of 26 minutes into hours before using the formula.

Now 26 minutes is simply 26/60 hours, which simplifies to 13/30 hours. We can write

$$v \;=\; \frac{22 \text{ miles}}{13/30 \text{ hours}}$$

This is a **complex fraction**: a fraction that has another fraction in the numerator, denominator, or both.

One way to calculate its value with a calculator is to use parentheses and input it as 22 ÷ (13 ÷ 30). Check out what happens if you input it as 22 ÷ 13 ÷ 30.

Tip: Instead of parentheses, you can use the reciprocal button ($\boxed{1/x}$) on your calculator. First calculate the value of the fraction inverted

(upside-down): $\dfrac{13/30}{22}$. This is the **reciprocal** of the fraction. You can input it as 13 ÷ 30 ÷ 22.

Once you've calculated the reciprocal, push the $\boxed{1/x}$ button to convert the reciprocal into the answer that you want.

---

8. Find the average speed in the given units.

   a. A duck flies 3 miles in 6 minutes.
   Give your answer in miles per hour.

   b. A lion runs 900 meters in 1 minute.
   Give your answer in kilometers per hour.

   c. Henry sleds 75 meters down the hill in 1.5 minutes.
   Give your answer in meters per second.

   d. Rachel swims 400 meters in 32 minutes.
   Give your answer in kilometers per hour.

9. Jake's grandparents live 150 km away from his home. One day it took him 2 h 14 min to get there and 1 h 55 min to come back home. In the questions below, round your answers to one decimal digit.

    a. What was his average speed going there?
       Hint: Change the time in hours and minutes into decimal hours.

    b. What was his average speed coming back?

    c. What was his overall average speed for the whole trip?

10. Another day Jake visited his grandparents again. Let's say that, because of the traffic, Jake achieved an average speed of 75 km/h going there but an average speed of only 65 km/h coming back.
How much longer did Jake spend driving home from his grandparents' place than going there?

How to remember the formula $d = vt$	Just solve for d from the formula $v = d/t$
I will show you how to *derive* that formula! Then you do not really have to memorize it. But you *do* need to remember the formula $v = d / t$. You can remember *that* formula with the trick I explained earlier—by thinking of the common unit for measuring speed (miles per hour): speed is miles per hour ↓ ↓ ↓ ↓ ↓ velocity = distance / time In symbols: $v = d / t$.	$v = \dfrac{d}{t}$ We want d alone, so we multiply both sides by t. $vt = \dfrac{d}{t} \cdot t$ The t's on the right side cancel. $vt = d$ We have it! Turning it around we get $d = vt$, which is the most common formula to show how velocity (v), time (t), and distance (d) are related.

11. Compare the average speeds to find which bird is faster: a seagull that flies 10 miles in 24 minutes or an eagle that flies 14 miles in half an hour?

12. A train normally travels at a speed of 120 km per hour. One day, the conditions were so icy and cold that it had to slow down to travel safely. So the train traveled the first half of its 160-km journey at half its normal speed. Then the weather improved, and the train was able to go faster again. It sped the remaining distance at twice its normal speed to make up time.

    a. How long did the train take to travel the whole distance (160 km)?

    b. How long would the train have taken if it had traveled the whole trip at its normal speed?

    c. What was the train's average speed for the trip on this cold and icy day?

13. How long will it take Charlotte to ride her bike from the music store to her home — a distance of 4.5 km — if she rides 1/3 of it at 12 km/h and the rest at 15 km/h?

The next problems are more challenging.

14. Your normal walking speed is 6 miles per hour. One day you walk slowly, at 3 miles per hour, half the distance from home to the swimming pool. Can you now make up for your slow walking by walking the remaining distance at double your normal speed?

    (Hint: Make up a distance between your home and the swimming pool for an example calculation. Choose an easy number.)

15. An airplane normally flies at a speed of 1,000 km/h. Due to some turbulence, it has to travel at a lower speed of 800 km/h for the first 40 minutes of a 1600-km trip. How fast should it fly for the rest of the trip so as to make up for the lost time?

    (Hint: You will also need to calculate the normal traveling time for this trip.)

Review 1

1. Solve. Check your solutions.

a. $\quad x + 7 \; = \; -6$	**b.** $\quad -x \; = \; 5 - 9$
c. $\quad 2 - x \; = \; -8$	**d.** $\quad 2 - 6 \; = \; -z + 5$
e. $\quad \dfrac{x}{11} \; = \; -12$	**f.** $\quad \dfrac{q}{-3} \; = \; -40$
g. $\quad 100 \; = \; \dfrac{c}{-10}$	**h.** $\quad \dfrac{a}{5} \; = \; -10 + (-11)$

Write an equation for the problem. Then solve it.

2. Alex bought three identical solar panels and paid a total of $837.
   How much did one cost?

   Equation:

Write an equation for the problem. Then solve it.

3. Andrew pays 1/7 of his salary in taxes. If he paid $187 in taxes, how much was his salary?

    Equation:

4. Use the formula $d = vt$ to solve the problem.

If you can bicycle at a speed of 20 km/h, how long will it take you to bicycle from the shopping center to a dentist's office, a distance of 1.2 km?	$d \;\; = \;\; v \quad t$ $\downarrow \quad\;\; \downarrow \;\; \downarrow$

5. Taking a bus, Emily can get to the community center that is 1.5 km from her home in 3 minutes. What is the average speed of the bus, in kilometers per hour?

6. Ed skates on his skateboard to school, which is 2 miles away. He travels half of the distance at a speed of 12 mph and the rest at a speed of 15 mph. How long does it take him to get to school?

Two-Step Equations

Just like the name says, **two-step equations take two steps to solve.** We need to apply two different operations to both sides of the equation. Study the examples carefully. It is not difficult at all!

Example 1. Solve $2x + 3 = -5$.

On the side of the unknown (left), there is a multiplication by 2 and an addition of 3. To isolate the unknown, we need to undo those operations.

$$
\begin{aligned}
2x + 3 &= -5 \quad &| -3 \\
2x &= -8 \quad &| \div 2 \\
x &= -4
\end{aligned}
$$

Check:

$$2 \cdot (-4) + 3 \overset{?}{=} -5$$

$$-8 + 3 \overset{?}{=} -5$$

$$-5 = -5 \ \checkmark$$

What if you divide first?

In this equation you *could* start by dividing by 2 and then subtract next. However, it is easier to subtract first, then divide, because that way you avoid dealing with fractions.

The solution below shows the steps if you divide by 2 first. Notice that the 3 on the left side also has to be divided by 2 to become 3/2.

$$
\begin{aligned}
2x + 3 &= -5 \quad &| \div 2 \\
x + (3/2) &= -5/2 \quad &| -3/2 \\
x &= -5/2 - 3/2 \\
x &= -4
\end{aligned}
$$

1. Solve. Check your solutions (as always!).

a. $\quad 5x + 2 \ = \ 67$	**b.** $\quad 3y - 2 \ = \ 71$
c. $\quad -2x + 11 \ = \ 75$	**d.** $\quad 8z - 2 \ = \ -98$

Example 2. In the equation below, the easiest thing to do is to multiply by 7 first, and then subtract.

$$\frac{x+2}{7} = 12 \qquad \Big|\cdot 7$$

$$\frac{7\cdot(x+2)}{7} = 84$$

$$x + 2 = 84 \qquad \Big|-2$$

$$x = 82$$

Check:

$$\frac{82+2}{7} \overset{?}{=} 12$$

$$\frac{84}{7} \overset{?}{=} 12$$

$$12 = 12 \quad \checkmark$$

2. Solve. Check your solutions (as always!).

a. $\dfrac{x+6}{5} = 14$

b. $\dfrac{x+2}{7} = -1$

c. $\dfrac{x-4}{12} = -3$

d. $\dfrac{x+1}{-5} = -21$

What if, in example 2, you subtract 2 first? (Optional)

It does not work. Subtracting 2 from both sides makes the equation more complicated.

$$\frac{x+2}{7} = 12 \qquad \Big|-2$$

$$\frac{x+2}{7} - 2 = 10$$

The quantity $(x + 2)$ is <u>also divided by 7.</u> It is $(x+2)/7$. Notice that, when calculating the value of this expression, the parentheses indicate that the addition is to be done first, and the division last. Therefore, when *undoing* the operations, we need to undo the division first.

However, it is possible to subtract first. But you need to subtract 2/7 instead of 2. To see that, let's write write the expression on the left side in a different way:

$$\frac{x+2}{7} = 12$$

$$\frac{x}{7} + \frac{2}{7} = 12 \qquad \Big|-2/7$$

$$\frac{x}{7} = 11\,5/7 \qquad \Big|\cdot 7$$

$$x = 82$$

Because it involves fractions, this solution is more complicated than the one shown in Example 2.

41

Example 3. Again, the unknown is "tangled up" with two different operations (division and addition), so to isolate it, we need two steps.

$$\frac{x}{4} + 5 = -2 \quad\bigg|\; -5$$

$$\frac{x}{4} = -7 \quad\bigg|\; \cdot\, 4$$

$$x = -28$$

Check:

$$\frac{-28}{4} + 5 \overset{?}{=} -2$$

$$-7 + 5 \overset{?}{=} -2$$

$$-2 = -2 \quad\checkmark$$

Solving it another way (optional) In case you wonder if we could multiply by 4 first, yes, in this case we can.

$$\frac{x}{4} + 5 = -2 \quad\bigg|\; \cdot\, 4$$

$$x + 20 = -8 \quad\bigg|\; -20$$

$$x = -28$$

Note: In the first step, <u>both</u> terms on the left side ($x/4$ and 5) have to be multiplied by 4!

It is a common student error to multiply only the first term of an expression by a number and to forget to multiply the other terms by that number.

3. Solve. Compare equation (a) to equation (b). They are similar, yet different! Make sure you know how to solve each one.

a. $\quad\dfrac{x}{10} + 3 = -2$

b. $\quad\dfrac{x+3}{10} = -2$

4. Solve. Compare equation (a) to equation (b). They are similar, yet different! Make sure you know how to solve each one.

a. $\quad\dfrac{x}{7} - 8 = -5$

b. $\quad\dfrac{x-8}{7} = -5$

Example 4. What's different about this one?   Check:

$$6 - 2n = 9 \qquad \Big|\; -6$$
$$-2n = 3 \qquad \Big|\; \div (-2)$$
$$n = -1\ 1/2$$

$$6 - 2 \cdot (-1\ 1/2) \overset{?}{=} 9$$
$$6 + 3 \qquad \overset{?}{=} 9 \quad \checkmark$$

5. Solve. Check your solutions (as always!).

a. $\quad 1 - 5x = 2$	**b.** $\quad 12 - 3y = -6$
c. $\quad 10 = 8 - 4y$	**d.** $\quad 7 = 5 - 3t$

6. Choose from the expressions at the right to build
   an equation that has the root $x = 2$.

$2x - 10$ $2x + 10$ 12
$5x + 6$
14 $3x - 9$
$3 \cdot 3$
$5x - 6$

7. Choose from the expressions at the right to build
   an equation that has the root $x = 5$.

$1 - 2x$ -4 $3x - 10$
$-8 \cdot 3$ -2 $-9 - 3x$
$-3x + 6$ $-2x - 1$

43

Example 5. Solve as a decimal.

$$\frac{3x}{7} = 0.9 \quad \Big| \cdot 7$$

$$\frac{7 \cdot 3x}{7} = 6.3$$

$$3x = 6.3 \quad \Big| \div 3$$

$$x = 2.1$$

Check:

$$\frac{3 \cdot 2.1}{7} \stackrel{?}{=} 0.9$$

$$\frac{6.3}{7} \stackrel{?}{=} 0.9$$

$$0.9 = 0.9 \checkmark$$

In this example, you *could* first divide by 3 and then multiply by 7. The solution wouldn't be any more difficult that way.

8. Solve.

a. $\dfrac{2x}{5} = 8$

b. $\dfrac{3x}{8} = -9$

c. $15 = \dfrac{-3x}{10}$

d. $2 - 4 = \dfrac{s + 4}{5}$

e. $\dfrac{x}{2} - (-16) = -5 \cdot 3$

f. $2 - 4p = -0.5$

44

Two-Step Equations: Practice

Example 1. The number line diagram illustrates the equation $$-29 + x + 7 = -4$$  Think of starting at −29, jumping x steps, jumping another 7 steps, and arriving at −4.	You can find the value of the unknown x using logical thinking or by writing an equation and solving it. Here is the solution using an equation. $$-29 + x + 7 = -4 \qquad \text{(add } -29 + 7 \text{ on the left side)}$$ $$-22 + x = -4 \qquad \vert + 22$$ $$x = 18$$

1. Write an equation to match the number line model and solve for the unknown.

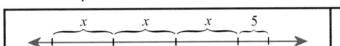

a.

b.

2. Solve. Compare equation (a) to equation (b). They are similar, yet different!

a. $\quad 2 - 5y = -11$	**b.** $\quad 5y - 2 = -11$

3. Solve. Compare equation (a) to equation (b). They are similar, yet different!

a. $\dfrac{2x}{7} = -5$	**b.** $\dfrac{x+2}{7} = -5$

4. Solve. Check your solutions (as always!).

a. $20 - 3y = 65$	**b.** $6z + 5 = -2.2$
c. $\dfrac{t+6}{-2} = -19$	**d.** $\dfrac{y}{6} - 3 = -0.7$

Example 2. The perimeter of an isosceles triangle is 26 inches and its base measures 5 inches. How long are the two sides that are equal (congruent)?

(To help yourself, label the image with the data from the problem.)

Read the two solutions below. Notice how neatly they tie in with each other!

Solution 1: an equation	**Solution 2: Logical thinking/mental math**
Let x *be* the unknown side length. We get: $$\text{perimeter} = x + x + 5$$ $$26 = 2x + 5$$ Next we solve the equation: $$2x + 5 = 26 \quad \mid -5$$ $$2x = 21 \quad \mid \div 2$$ $$x = 10.5$$ The two other sides measure 10.5 inches each.	The perimeter is 26 inches. This means that the two unknown sides and the 5-inch side add up to 26 inches. Therefore, if we subtract the base side, the two unknown sides must add up to 21 inches. So one side is half of that, or 10.5 inches.

5. Solve each problem below in two ways: write an equation, and use logical reasoning/mental math.

a. A quadrilateral has three congruent sides. The fourth side measures 1.4 m.
   If the perimeter of the quadrilateral is 7.1 meters, what is the length of each congruent side?

Equation:	Mental math/logical thinking:

b. You bought six identical baskets from an artisan. She gave you a $12 discount on your order, and your total bill was only $46.80. What is the normal price of one basket?

Equation:	Logical thinking:

c. Two-fifths of a number is 466. What is the number?

Equation:	Logical thinking:

Growing Patterns

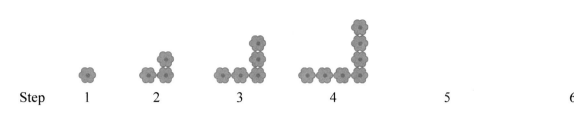

Step   1   2   3   4   5   6

How do you think this pattern is growing?

How many flowers will there be in step 39?

This pattern adds 2 flowers in each step, except in step 1. This means that by step 39, we have added 2 flowers 38 times. Therefore, there are $1 + 2 \cdot 38 = 77$ flowers in step 39.

Write a formula for the number of flowers in step *n*.

There are several ways to do this. The three ways explained below are not the only ones!

1. Let's view the pattern as adding 2 flowers in each step after the first one. By step *n*, the pattern has added one less than *n* times 2 flowers, because we need to exclude that first step. This means that $(n - 1)$ times 2, or $(n - 1) \cdot 2$, flowers added to the one flower that we started with.

   This gives us the expression $1 + (n - 1) \cdot 2$. Since we customarily put the variable first and the constant last, we can rewrite that expression as $1 + 2(n - 1)$ and then as $2(n - 1) + 1$.

2. Another way to think about this pattern is as two legs. One leg includes the flower in the corner, so it has the same number of flowers as the step number. The other leg doesn't have the corner flower, so it has one flower less than the step numbers. In other words, in step 3, we have $3 + 2$ flowers. In step 4, we have $4 + 3$ flowers. In step 5, we have $5 + 4$ flowers.

   This gives us a formula for the number of flowers in step *n*: there are $n + (n - 1)$ flowers in step *n*.

3. Yet another way is that, in each step, there are twice as many flowers as the step number, minus one for the flower that is shared. For example, in step 4, we have twice 4 minus 1, which is seven flowers.

   This also gives us a formula: there are $2n - 1$ flowers in step *n*.

All of the formulas are equivalent (just as we would expect!) and simply represent different ways of thinking about the number of flowers in each step. On the right, you can see how the first two formulas can be simplified to the third one.

$$
\begin{array}{ll}
n + (n - 1) & 2(n - 1) + 1 \\
= n + n - 1 & = 2n - 2 + 1 \\
= 2n - 1 & = 2n - 1
\end{array}
$$

In which step are there 583 flowers?

We can use our formula to write an equation to answer this question. In the question the step number *n* is unknown, but the total number of flowers in that step is 583. Since we know from our formula that there are $2n - 1$ flowers in step *n*, we get

$$
\begin{array}{rcl|l}
2n - 1 & = & 583 & +1 \\
2n & = & 584 & \div 2 \\
n & = & 292 &
\end{array}
$$

49

1　　　　　　　　2　　　　　　　　3　　　　　　　　4　　　　　　　　5

1. **a.** How is this pattern growing?

   b. How many triangles will there be in step 39?

   c. Write a formula for the number of triangles in step *n*.
      Check your answer with your teacher before going on to part (d).

   d. In which step will there be 311 triangles?
      Write an equation and solve it.
      Notice, this question is <u>different</u> from the one in part (c).

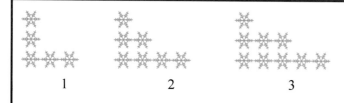

　　1　　　　　　　2　　　　　　　　3　　　　　　　　4　　　　　　　　5

2. **a.** How do you think this pattern is growing?

   b. How many snowflakes will there be in step 39?

   c. Write a formula for the number of snowflakes in step *n*.
      Check your answer with your teacher before going on to part (d).

   d. In which step will there be 301 snowflakes?
      Write an equation and solve it.

	Instead of showing the steps of the pattern horizontally, like this...	...we can also show them like this:

Now, each row of flowers is one step of the pattern.

---

3. A section of a flower garden has rows of flowers. The first row has four flowers, and each row after that has one more flower than the previous row.

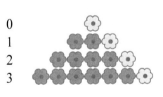

Row	Flowers
1	3 + 1
2	3 + 2
3	3 + 3
4	3 + 4
n	

a. Write a formula that tells the gardener the number of flowers in row n.

b. How many flowers are in the 28th row?

c. In which row will there be 97 flowers? Write an equation and solve it.

4. This pattern is similar to the previous one. This time each row has *two* more flowers than the previous row. Notice that the number of flowers in each row gives us the list of all the odd numbers.

Here's one way to look at this pattern: Label the first row as "row 0." Then the number of flowers in any row is twice the row number plus 1.

a. Write a formula that tells the gardener the number of flowers in row n.

b. In which row will there be 97 flowers? Write an equation and solve it.

5. Each floor of a multi-story building is 9 feet high.

    a. Write an expression for the total height of the building, if it is *n* stories high.

    b. Write an expression for the total height of the building if it is *n* stories high, and the bottom of the first floor is elevated 2 ft above the ground.

    c. How many stories does the building have if the total height of the building is 164 ft? Solve this problem in two ways: using logical thinking and using an equation.

6. Jeremy earns $400 a week. He also earns $15 for every hour he works overtime.

    a. Write an expression for Jeremy's *total* earnings if he works *n* hours overtime. You can use the table to help you.

    b. How many hours should Jeremy work overtime in order to earn $970 in a week?

Overtime hours	Total earnings
0	
1	
2	
17	
n	

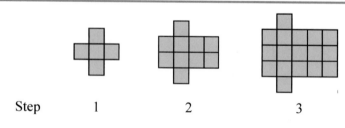

| Step | 1 | 2 | 3 | 4 | 5 |

What is the pattern of growth here?
How many squares will there be in step 59?

Puzzle Corner

52

A Variable on Both Sides

Example 1. Solve $2x + 8 = -5x$.

Notice that the unknown appears on both sides of the equation. To isolate it, we need to

- ¿ either subtract $2x$ from both sides—because that makes $2x$ disappear from the left side

- ¿ or add $5x$ to both sides—because that makes $-5x$ disappear from the right side.

$$
\begin{array}{rl}
2x + 8 = -5x & \quad |+5x \\
2x + 8 + 5x = 0 & \quad \text{(now add } 2x \text{ and} \\
& \quad \quad 5x \text{ on the left side)} \\
7x + 8 = 0 & \quad |-8 \\
7x = -8 & \quad |\div 7 \\
x = -8/7 &
\end{array}
$$

Check:

$$2 \cdot (-8/7) + 8 \overset{?}{=} -5 \cdot (-8/7)$$

$$-16/7 + 8 \overset{?}{=} 40/7$$

$$-2\,2/7 + 8 = 5\,5/7 \quad \checkmark$$

Example 2. Solve $10 - 2s = 4s + 9$.

To isolate s, we need to

- ¿ either add $2s$ to both sides

- ¿ or subtract $4s$ from both sides.

The choice is yours. Personally, I like to keep the unknown on the left side and eliminate it from the right.

$$
\begin{array}{rl}
10 - 2s = 4s + 9 & \quad |-4s \\
10 - 2s - 4s = 9 & \quad \text{(now simplify } -2s - 4s \\
& \quad \quad \text{on the left side)} \\
10 - 6s = 9 & \quad |-10 \\
-6s = -1 & \quad |\div(-6) \\
s = 1/6 &
\end{array}
$$

Check:

$$10 - 2 \cdot (1/6) \overset{?}{=} 4 \cdot (1/6) + 9$$

$$10 - 2/6 \overset{?}{=} 4/6 + 9$$

$$9\,4/6 = 9\,4/6 \quad \checkmark$$

1. Solve. Check your solutions (as always!).

a. $\quad 3x + 2 = 2x - 7$	**b.** $\quad 9y - 2 = 7y + 5$

2. Solve. Check your solutions (as always!).

a. $11 - 2q \;\; = \;\; 7 - 5q$	**b.** $6z - 5 \;\; = \;\; 9 - 2z$
c. $8x - 12 \;\; = \;\; -1 - 3x$	**d.** $-2y - 6 \;\; = \;\; 20 + 6y$
e. $6w - 6.5 \;\; = \;\; 2w - 1$	**f.** $5g - 5 \;\; = \;\; -20 - 2g$

Combining like terms

Remember, in algebra, a *term* is an expression that consists of numbers, fractions, and/or variables that are multiplied. This means that the expression $-2y + 7 + 8y$ has three terms, separated by the plus signs.

In the expression $-2y + 7 + 8y$, the terms $-2y$ and $8y$ are called **like terms** because they have the same variable part (in this case a single y). We can **combine** (add or subtract) like terms.

To do that, it helps to organize the terms in the expression in alphabetical order according to the variable part and write the constant terms last. We get $-2y + 8y + 7$ ($8y - 2y + 7$ is correct, too).

Next, we add $-2y + 8y$ and get $6y$. So the expression $-2y + 7 + 8y$ simplifies to $6y + 7$.

Example 3. Simplify $6y - 8 - 9y + 2 - 7y$.

First, we organize the expression so that the terms with y are written first, followed by the constant terms.

For that purpose, we **view each operation symbol (+ or −) in front of the term as the *sign* of each term.** In a sense, you can imagine each plus or minus symbol as being "glued" to the term that follows it. Of course the first term, $6y$, gets a "+" sign.

After reordering the terms, the expression becomes $6y - 9y - 7y - 8 + 2$.

Now we need to combine the like terms $6y$, $-9y$, and $-7y$. We do that by finding the sum of their coefficients 6, −9, and −7. Since $6 - 9 - 7 = -10$, we know that $6y - 9y - 7y = -10y$.

Similarly, we combine the two constant terms: $-8 + 2 = -6$.

Our expression therefore simplifies to $-10y - 6$.

Why can we do it this way?

Because subtracting a term is the same as adding its opposite. In symbols,

$$6y \quad -8 \quad -9y +2 \quad -7y$$
$$= 6y + (-8) + (-9y) + 2 + (-7y).$$

In other words, the expression $6y - 8 - 9y + 2 - 7y$ is the SUM of the terms $6y$, -8, $-9y$, 2, and $-7y$.

---

3. Fill in the pyramid! Add each pair of terms in neighboring blocks and write its sum in the block above it.

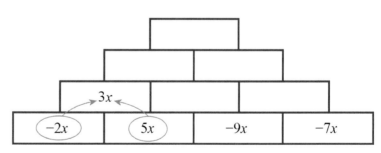

4. Organize the expressions so that the variable terms are written first, followed by constant terms.

  a. $6 + 2x - 3x - 7 + 11$

  b. $-s - 12 + 15s + 9 - 7s$

  c. $-8 + 5t - 2 - 6t$

5. Simplify the expressions in the previous exercise.

6. Simplify.

a. $5x - 8 - 7x + 1$	**b.** $-6a - 15a + 9a + 7a$
c. $-8 + 7c - 11c + 8 - c$	**d.** $10 - 5x - 8x - 9 + x$

7. Fill in the pyramid! Add each pair of terms
   in neighboring blocks and write its sum
   in the block above it.

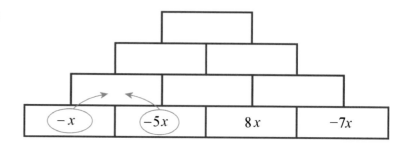

8. Find what is missing from the sums.

    a. $8x + 2 +$ _____ $= 5x + 8$ **b.** $5b - 2 +$ _____ $= 2b + 7$

    c. $-2z +$ _____ $= 1 - 5z$ **d.** $-4f + 3 +$ _____ $= -f - 1$

9. Fill in the pyramid! Add each pair of terms
   in neighboring blocks and write its sum
   in the block above it.

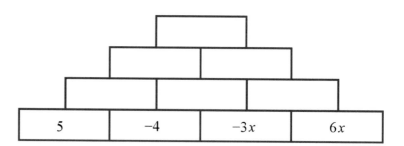

10. Simplify.

a. $0.5y + 1.2y - 0.6y$	**b.** $-1.6v - 1 - v$	**c.** $-0.8k + 3 + 0.9k$

11. A challenge! Solve the equation $(-1/2)x - 6 + 8x + 7 - x = 0$.

Example 4. One or both sides of an equation may have several terms with the unknown. In that case, we need to combine the like terms (simplify) before continuing with the actual solution.	$3x + 7 - 5x = 6x + 1 - 5x$ $-2x + 7 = x + 1 \quad \vert \; -x$ $-3x + 7 = 1 \quad \vert \; -7$ $-3x = -6 \quad \vert \; \cdot(-1)$ $3x = 6 \quad \vert \; \div 3$ $x = 2$	On the left side, combine $3x$ and $-5x$. On the right side, combine $6x$ and $-5x$.

12. Solve. Check your solutions.

a. $\quad 6x + 3x + 1 = 9x - 2x - 7$	b. $\quad 16y - 4y - 3 = -4y - y$
c. $\quad -26x + 12x = -18x + 8x - 6$	d. $\quad -9h + 4h + 7 = -2 + 5h + 9h + 8h$

13. Solve. Check your solutions.

a. $\quad 2x - 4 - 7x \;=\; -8x + 5 + 2x$	**b.** $\quad -6 - 4z - 3z \;=\; 5z + 8 - z$
c. $\quad 8 - 2m + 5m - 8m \;=\; 20 - m + 5m - 2m$	**d.** $\quad -x - x + 2x \;=\; 5 - 5x + 9x$
e. $\quad -q + 2q - 5q - 6q \;=\; 20 - 7 - 9 + q$	**f.** $\quad 9 - s + 7 - 9s \;=\; 2 - 2s - 11$

Some Problem Solving

Ratio problems and equations

Example 1. A truck is carrying 21 pallets of beans. Each pallet weighs 2,400 pounds and contains bags of pinto beans, navy beans, and black beans in a ratio of 7:3:2. Calculate the total weight of each type of beans in the truck.

We can solve this using a bar model:

24,000 lb

There are $7 + 3 + 2 = 12$ parts in total. The weight of one part is $24,000 ÷ 12 = 2,000$ lb.

Or we can write an equation. The unknown x is the same as in the bar model: it is one part.

$$7x + 3x + 2x = 24,000$$
$$12x = 24,000 \qquad \Big| ÷ 12$$
$$x = 2,000$$

But the weight of one part is not the final answer! We need to calculate the weight of each type of beans:

¿ pinto beans: $21 · 7 · 2,000$ lb $= 294,000$ lb

¿ navy beans: $21 · 3 · 2,000$ lb $= 126,000$ lb

¿ black beans: $21 · 2 · 2,000$ lb $= 84,000$ lb

1. Jaime and Juan divided a profit of \$820 in a ratio of 2:3. Calculate each one's part.

Jaime:

Juan:

2. The *aspect ratio* of a photograph (the ratio of its width to its height) is 4:7 and its perimeter is 66 cm. How long and how wide is the photograph?

Solve this problem in two ways, and observe how the two solutions relate to each other.

a. Solve the problem using logical reasoning.

b. Solve the problem using an equation.

3. You bought five computer keyboards from an online store. A shipping charge of $8.75 is included in the final bill, which comes to $78.20. How much did one keyboard cost?

    a. Solve the problem using logical reasoning.

    b. Write an equation for the problem, and solve it. Notice how the solution steps of the equation correspond to the steps of solving the problem using logical thinking.

4. Computer mice are on sale for $12 each from an online store. If your order is under $250, the store charges a shipping and handling fee of $7.90. You buy as many mice as you can with $100. How many mice can you buy?

    a. Solve the problem using logical reasoning.

    b. Write an equation for the problem, and solve it. Notice how the solution steps of the equation correspond to the steps of solving the problem using logical thinking.

5. The perimeter of the rectangle is 76 cm.
   Find its length and width.

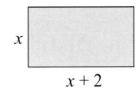

x

$x + 2$

6. **a.** Sketch an equilateral triangle with a perimeter of $24x + 9$.

   b. Sketch a rectangle with a perimeter of $20x + 6$.

7. **a.** Write an equation to solve for the unknown a.

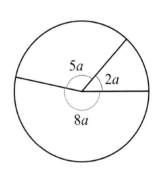

5a 2a 8a

   b. Find the measure in degrees of each of the three angles.

8. Four adjacent (side by side) angles form the line l.

   a. Write an equation to solve for the unknown a.

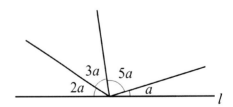

3a 5a
2a a
l

   b. Find the measure of each of the four angles, rounded to the nearest 0.01 degree.

Using The Distributive Property

Sometimes you may have to use the distributive property to remove parentheses before solving an equation.	**Example 1.**

Example 1.

$$2(x + 9) = -2x$$
$$2x + 18 = -2x \qquad \mid +2x$$
$$4x + 18 = 0 \qquad \mid -18$$
$$4x = -18 \qquad \mid \div 4$$
$$x = -4.5$$

1. Solve.

a. $5(x + 2) = 85$

b. $9(y - 2) = 6y$

c. $2(x + 7) = 6x - 11$

d. $-3z = 5(z + 9)$

e. $10(x + 9) = -x - 5$

f. $2(5 - s) = 3s - 1$

Example 2. Sometimes you can start the solution to an equation, such as $3(x + 5) = -18$, in two different ways.

Divide by a common factor:	$3(x + 5) = -18$ $\quad\bigg\| \div 3$ $x + 5 = -6$ $\quad\bigg\| -5$ $x = -11$	Distribute the multiplication:	$3(x + 5) = -18$ $3x + 15 = -18$ $\quad\bigg\| -15$ $3x = -33$ $\quad\bigg\| \div 3$ $x = -11$

2. Solve in two ways: (i) by dividing first and (ii) by distributing the multiplication over the parentheses first.

a. $6(x - 7) = 72$

Divide first:	Distribute the multiplication first:

b. $10(q - 5) = -60$

Divide first:	Distribute the multiplication first:

3. What is to be done in each step? Fill in the missing operations to be done to both sides of the equation.

a.
$$2(x + 9) = -2x$$
$$x + 9 = -x$$
$$2x + 9 = 0$$
$$2x = -9$$
$$x = -4\,1/2$$

b.
$$\frac{x + 2}{7} = 6x$$
$$x + 2 = 42x$$
$$2 = 41x$$
$$41x = 2$$
$$x = 2/41$$

Switch the sides.

4. Solve.

a. $2(x + 7) = 3(x - 6)$	**b.** $5(y - 2) + 6 = 9$
c. $2x - 1 = 3(x - 1)$	**d.** $3(w - \frac{1}{2}) = 6(w + \frac{1}{2})$
e. $20(x - \frac{1}{4}) = x - 2$	**f.** $8 + 2(7 - v) = 13$

5. Select from the expressions at the right to make an equation that has -3 as its root.

$-x + 5$ $x - 5$ $4x$

12 $2(x + 4)$

$6(x + 1)$ $2(x - 4)$ $6(x - 1)$

6. Write an equation for this number line diagram, and find the value of the unknown a.

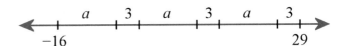

7. Draw a number line diagram for each equation, and solve.

   a. $-5 + 3x = 16$

   b. $-5 + 3(x + 2) = 16$

The distributive property works the same way with negative numbers as with positive ones. As you practice more, you'll be able to start doing these multiplications in a single step instead of several.

Example 3.

$$-2(3x + 7) = -2 \cdot 3x + (-2) \cdot 7$$

$$= -6x + (-14)$$

$$= -6x - 14$$

We write $+ (-14)$ as $- 14$, because that is simpler.

Example 4.

$$-9(3s - 3)$$

$$= -9 \cdot 3s - 9(-3)$$

$$= -27s - (-27)$$

$$= -27s + 27$$

The expression $9(-3)$ means $9 \cdot (-3)$. Similarly, $9 \cdot 3$ could be written as $(9)(3)$.

8. Use the distributive property to multiply.

a. $-11(x + 2) =$	**b.** $-3(y + 5) =$	**c.** $-7(x - 2) =$
d. $-3(y - 20) =$	**e.** $-10(2x + 10) =$	**f.** $-3(2y + 90) =$
g. $-0.5(w + 16) =$	**h.** $-0.8(5x - 20) =$	**i.** $-200(0.9x + 0.4) =$

9. Solve.

a. $\quad -2(x + 5) \quad = \quad 5x$	**b.** $\quad -6(y - 2) \quad = \quad 3(y + 2)$
c. $\quad -0.1(40y + 90) \quad = \quad 2$	**d.** $\quad 3 - t \quad = \quad -5(t - \frac{1}{2})$

Example 5. The expression $-(x + 5)$ simply means $-1(x + 5)$.
We can simplify it using the distributive property (on the right) →

$$-1(x + 5) = -1x + (-1)(5)$$
$$= -x + (-5)$$
$$= -x - 5$$

This is an example of an important principle in algebra:

If you multiply any number by −1, you get the opposite of that number. In symbols,

$$-1x = -x$$

So the expression $-(x + 5)$ means the *opposite* of $x + 5$. To simplify it, we take the opposite of each term in the quantity (the opposite of *x* *and* the opposite of 5) to get $-x - 5$.

Example 6. Simplify $-(2s - 5)$.

We multiply both terms in the quantity $2s - 5$ by -1. This ends up **changing the sign of each term in the quantity**, because each term becomes its opposite when multiplied by -1. So, both $2s$ and -5 change their signs and become $-2s$ and $+5$.

$$-(2s - 5) =$$
$$(-1)2s - (-1)5$$
$$= -2s + 5$$

10. Simplify.

a. $-(x + 4) =$	**b.** $-(y - 9) =$	**c.** $-(x - 12) =$
d. $-3(a - 6) =$	**e.** $2(-x + 5) =$	**f.** $-4(x - 1.2) =$
g. $-(4y - 20) =$	**h.** $-(5t + 0.9) =$	**i.** $-(2y + 1.4) =$
j. $-7(0.2 - w) =$	**k.** $0.1(-60t - 7) =$	**l.** $-0.2(9y + 4) =$

11. Here's another riddle to discover by solving the equations.

A. $3(2x - 4) = 3x$	**E.** $-2(x + 4) = 11$	**A.** $-(x + 4) = 10 + x$
S. $-2x + 4 = 3(5 - x)$	**A.** $-0.1(50x - 5) = x$	**R.** $2 = -2(x + 4)$
L. $5x = -(-2x - 3)$	**D.** $-6(3x + 1) = 1 + x$	**Y.** $3(2x + 5) = 4x + 5 + x$

Where can you buy a ruler that is 3 feet long?	At...		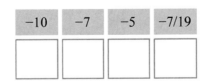				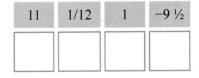			

 4

 −10 −7 −5 −7/19

 11 1/12 1 −9 ½

The root of each equation below is 2. Find each of the missing numbers.

a. 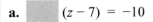 $(z - 7) = -10$ **b.** $(x - 5) = 3x + 6$

c. $(y + 8) = 2y - 6$ **d.** $w = -6(w + 1)$

Puzzle Corner

Word Problems 2

1. Which equation matches the problem? Once you find it, solve the equation.

    a. Mrs. Hendrickson bought herself a cup of coffee that cost $3. She
also bought ice cream cones that cost $2.20 each for each of her
preschoolers. She paid for all of it with $25. How many ice cream
cones did she buy?

$n(3 + 2.20) = 25$	$3 + 2.20n = 25$
$3(n + 2.20) = 25$	$3n + 2.20 = 25$

    b. Mr. Sanchez spent exactly $35 to treat some people in his bicycling
club to a cup of coffee and an ice cream cone each. Each coffee cost
$3, and each ice cream cone cost $2.83. How many people were
treated to coffee and ice cream by Mr. Sanchez?

$n(3 + 2.83) = 35$	$3 + 2.83n = 35$
$3(n + 2.83) = 35$	$3n + 2.83 = 35$

2. Write an equation for the following problem and solve it.
   The perimeter of a rectangle is 144 cm. Its length is 28 cm. What is its width?

3. An online store sells herbal sleeping aids for $12 per bottle. A fixed shipping and handling (S&H) fee of $5 is added to each order. (The S&H fee is not per bottle, but per order.)

    a. Write an expression for the total cost of buying five bottles.

    b. Write an expression for the total cost of buying n bottles.

    c. Write an equation for this question, and solve it.
       How many bottles can Mrs. Brooks get with $173?

4. Let's change the shipping and handling fee to be more realistic. Each bottle still costs $12, but the shipping and handling fee is $5 for the first bottle you buy, and $1 for each additional bottle in the same order.

    a. Write an expression for the total cost of buying five bottles.

    b. Write an expression for the total cost of buying eight bottles.

    c. Write an expression for the total cost of buying n bottles (where n is more than 1).

    d. Write an equation for this question, and solve it.
       How many bottles can Mrs. Brooks get with $173?

5. Based on the cost in problem 4, how many bottles can Mrs. Brooks buy with $300?

    a. Solve the problem without an equation.

    b. Solve this problem using an equation. Compare the solution steps to how you solved it in (a).

6. The total area of a sandbox with two compartments is 8 square meters. Solve for x.

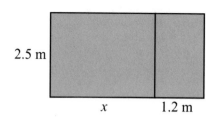

7. Before moving to a different suburb, Tony used to commute 14 km from his home to his workplace every workday. After his move, he was going to measure the new distance to his workplace using his car's odometer but forgot to check it until he had gone to work and back home three times. By then it showed 123.6 km.

   How much longer is Tony's commute now than it was in the past?

   a. Solve the problem using logical reasoning and arithmetic.

   b. Write an equation for the problem, and solve it.

8. A school schedule has six class periods of equal length plus a 30-minute study hall each day. The total time each day that students spend in classes and study hall is 360 minutes. How long is each class period?

   a. Solve the problem using logical reasoning and arithmetic.

   b. Solve the problem by writing an equation. Compare the steps of solving the equation to those in your method in part (a).

Inequalities

An **inequality** contains two expressions that are separated by one of these signs $<$, $>$, $\leq$ or $\geq$.

(expression 1) $<$ (expression 2)

The sign $\leq$ is read, "less than or equal to." It is the $<$ sign and $=$ sign combined.
The sign $\geq$ is read, "greater than or equal to." It is the $>$ sign and $=$ sign combined.

For example, the inequality $6y \geq -2$ is read, "$6y$ is greater than or equal to a negative two."

Typically, an inequality has an infinite number of solutions. The set of all the possible solutions to an inequality is its **solution set**. We can **plot the solution set of an inequality on a number line.**

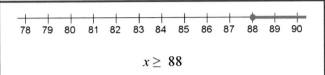

$x \geq 88$

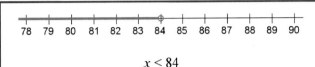

$x < 84$

We draw a *closed* circle at 88 because 88 fulfills the inequality. Any number that is greater than 88 (such as 88.1 or 90 1/2) also fulfills the inequality.

We draw an *open* circle at 84, because 84 *does not* fulfill the inequality ($84 < 84$ is a false statement). Any number that is less than 84 works fine!

To solve an inequality, we can

- ¿ add the same number to both sides
- ¿ subtract the same number from both sides
- ¿ multiply or divide both sides by the same number.

So solving inequalities works essentially the same way as solving equations, but there is *one* exception. That is, if you divide or multiply an inequality by a *negative* number, you need to reverse the sign of the inequality (from $<$ to $>$, or $\leq$ to $\geq$, and vice versa). For example, multiplying the inequality $-7 < 7$ by -1 yields the inequality $7 > -7$. However, we will not be dealing with the exception in this book. You will be solving only inequalities where you multiply or divide the inequality by a *positive* number.

Example 1.
$$5x + 3 \; < \; 6 \qquad |-3$$
$$5x \; < \; 3 \qquad |\div 5$$
$$x \; < \; 3/5$$

Example 2.
$$\frac{t-20}{7} \; \geq \; 7 \qquad |\cdot 7$$
$$\frac{\cancel{7} \cdot (t-20)}{\cancel{7}} \; \geq \; 49$$
$$t - 20 \; \geq \; 49 \qquad |+20$$
$$t \; \geq \; 69$$

1. Write an inequality that corresponds to the plot on the number line.

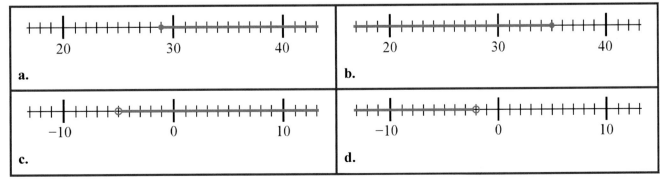

a.

b.

c.

d.

> **Example 3.** Mom said, "Don't spend more than $100." "Not more than" means "less than or equal to."
> You can spend $100 or any amount less than $100, but you cannot spend $100.29. We can represent
> this statement as
>
> $$money\ spent \leq \$100$$
>
> Using the variable m for the "money spent," we can write the inequality $m \leq \$100$.
>
> ---
>
> The symbol $\geq$ (greater than or equal to) often corresponds to the phrase "at least."
>
> "Could you get me at least 20 apples?" Using n for the number of apples, we can write $n \geq 20$.

2. Write an inequality for each phrase. Choose a variable to represent the quantity in question.

   a. You have to be at least 21 years of age.

   b. He has been to at least a dozen countries.

   c. Citizens 55 years or older get a free admission.

   d. We did not see more than 12 birds.

   e. I need at least 50 screws for the project.

3. Make up a situation from real life that could be described by the given inequality.

   a. $a < 30$

   b. $p > 100$

   c. $b \geq 8$

   d. $x \leq 60$

> **Example 4.** Solve the inequality $4x - 1 < 11$ in the set $\{1, 2, 5, 7, 9, 11\}$.
>
> This simply means to try each number from the set to see if it fulfills the inequality.
> For example, let's try 11. Is $4 \cdot 11 - 1$ less than 11? No, it's not (it's 43). So 11 is not a solution.
>
> In this case, the numbers 1 and 2 are the only solutions to the inequality. The solution set
> of this inequality is thus $\{1, 2\}$.

4. **a.** Solve the inequality $7x - 13 < 45$ in the set $\{1, 3, 5, 7, 9, 11\}$.

   b. Solve the inequality $3x + 10 \geq 25$ in the set $\{2, 4, 6, 8, 10\}$.

   c. Solve the inequality $2 - y \geq y + 1$ in the set $\{-3, -2, -1, 0, 1, 2, 3\}$.

5. Solve these inequalities by applying the same operation to both sides. Notice that the inequality symbol ($<$, $>$, $\leq$ or $\geq$) does not change.

a.	b.	c.
$3y \quad < \quad 48$	$y - 8 \quad > \quad 59$	$2c - 5 \quad \geq \quad 23$
$<$	$>$	$\geq$
		$\geq$

6. Solve the inequalities and plot their solution sets on a number line. Write appropriate multiples of ten under the bolded tick marks (for example, 30, 40, and 50).

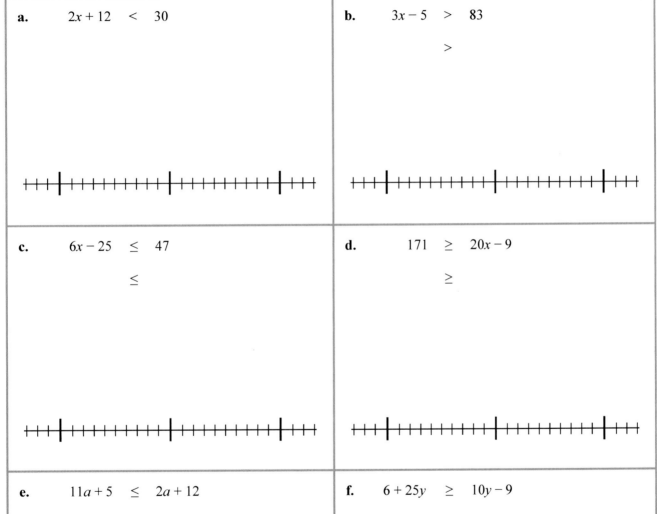

a. $\quad 2x + 12 \quad < \quad 30$

b. $\quad 3x - 5 \quad > \quad 83$

$\quad >$

c. $\quad 6x - 25 \quad \leq \quad 47$

$\quad \leq$

d. $\quad 171 \quad \geq \quad 20x - 9$

$\quad \geq$

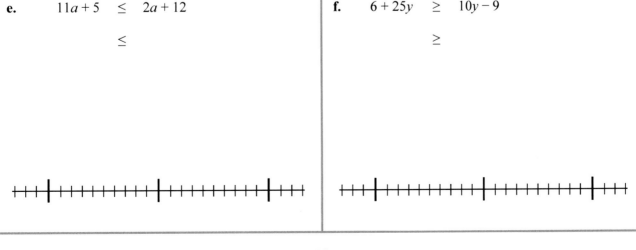

e. $\quad 11a + 5 \quad \leq \quad 2a + 12$

$\quad \leq$

f. $\quad 6 + 25y \quad \geq \quad 10y - 9$

$\quad \geq$

73

How do you check the solution to an inequality? After all, there is an infinite number of solutions.

To check the solution to an inequality, choose some numbers from the solution set and check that they do indeed fulfill the inequality. Then choose some numbers that are *not* in the solution set and check that they *don't* fulfill the inequality.

$$\textbf{Example.} \quad 6x - 2 > 3 \qquad | + 2$$
$$6x > 5 \qquad | \div 6$$
$$x > 5/6$$

To check that the solution $x > 5/6$ is correct, choose a few numbers that are greater than 5/6 and substitute them into the original inequality. It is best to choose at least one number that is near 5/6. Let's choose 1 and 2. The inequality should be true.

$$6 \cdot 1 - 2 \overset{?}{>} 3 \qquad\qquad\qquad 6 \cdot 2 - 2 \overset{?}{>} 3$$
$$4 > 3 \checkmark \qquad\qquad\qquad\qquad 10 > 3 \checkmark$$

Then, we choose some numbers that are *not* in the solution set, say 0 and −5. This time, the inequality should NOT hold true.

$$6 \cdot 0 - 2 \overset{?}{>} 3 \qquad\qquad\qquad 6 \cdot (-5) - 2 \overset{?}{>} 3$$
$$-2 > 3 \ \times \qquad\qquad\qquad\qquad -32 > 3 \ \times$$

Our check is complete and everything looks fine!

7. Solve the inequalities. Check your solution using the method above.

a. $\quad 2x - 3 > -9$	**b.** $\quad 9 + 3x \leq 30$
c. $\quad 5 + 10x < 22$	**d.** $\quad -20 + 3x \leq 19$

8. Which inequality has its solution plotted on the number line?

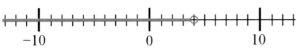

a.	**(i)**	$7x - 5 \ < \ 3$	**b.**	**(i)**	$4x - 10 \ \leq \ -34$
	(ii)	$5x - 3 \ < \ 7$		**(ii)**	$4x + 10 \ \geq \ -34$
	(iii)	$3x - 7 \ < \ 5$		**(iii)**	$4x - 10 \ \geq \ -34$

9. Solve the inequalities. Plot the solution set on the number line. Do not forget to check your answers.

a. $5x - 17 \ < \ 43$

b. $20x + 18 \ > \ 258$

The set of **natural numbers** or **counting numbers** is {1, 2, 3, 4, 5, 6...}.

10. **a.** What solutions does the inequality $x < 6$ have in the set of natural numbers?

b. What solutions does the inequality $y > 11$ have in the set of *even* natural numbers?

c. Find the natural numbers n that fulfill both inequalities: $n > 12$ and $n + 1 < 21$

11. Find three integers that fulfill both inequalities: $15y - 12 < 20$ and $2y + 6 > -5$.

Puzzle Corner

Find a number to fill in the empty space so that the solution to the inequality
☐$x - 7 < 23$ will be $x < 5$.

Word Problems and Inequalities

Example. As a salesperson, you are paid $150 per week plus $12 per sale. This week you want to earn at least $250. Write an inequality for the number of sales you need to make, solve it, and describe the solutions.

Let n be the number of sales, as that is the unknown. Each week, you earn $150 + $12 \cdot n$. Let's write this expression without the dollar symbols and in the usual order of terms: $12n + 150$.

That expression, with the condition that you want to earn at least $250, gives the inequality $12n + 150 \geq 250$.

$$12n + 150 \geq 250 \qquad | - 150$$
$$12n \geq 100 \qquad | \div 12$$
$$n \geq 8\ 4/12$$
$$n \geq 8\ 1/3$$

Since n is the number of sales, it has to be a whole number. Therefore, if you make 9 or more sales, you will be paid at least $250.

Another way to solve this is to think in terms of the amount you want to earn on top of your $150 base pay. In that case, you can write the inequality $12n \geq 100$.

1. Let n represent the amount of sales you make as a salesperson. Which description(s) of the solution set are correct?

a. $n < 18$	**b.** $n \leq 30$
You make at least 18 sales.	You make at least 30 sales.
You make at most 17 sales.	You make 29 sales or fewer.
You make at most 18 sales.	You make at most 30 sales.
You make 19 sales or more.	You make 31 sales or more.

2. Jeannie earns $350 per week plus $18 for each hour of overtime that she works. How many hours of overtime does she need to work if she wants to earn at least $500?

Write an inequality and solve it. Plot the solution set on a number line.
Lastly, explain the solution set in words.

Write an inequality for each problem, and solve it. Plot the solution set on a number line.
Lastly, explain the solution set in words.

3. Sheila is packing decorative candles for shipping. Each candle comes in its own box, which is 6.5 cm tall. The carton for shipping them is 45 cm tall. If she begins packing by putting a 1-cm-thick pad in the bottom, then how many boxes of candles will she be able to stack vertically in the carton for shipping?

4. You have a coupon that gives you a $10 discount on your total order at SuperSales store. How many pairs of socks can you buy, if you have only $55 in cash to spend, and the socks cost $6.70 a pair?

5. Joey has to work a minimum of 160 hours as an apprentice in a print shop before he can be hired permanently. He has already completed 21 hours of training. Now he has an opportunity to work 7.5-hour workdays. How many days will he need to work in order to finish at least the minimum training?

6. You want to print an e-book at a local office center. Printing black and white pages costs $0.03 per page, and then there is a fixed binding fee of $2.45.

How many pages can you print for no more than $35?

Graphing

Do you remember equations with two variables? When an equation has two variables (like the equation $y = 2x - 3$), it usually has an infinite number of solutions. In other words, there is an infinite number of values for x and y that make the equation true.

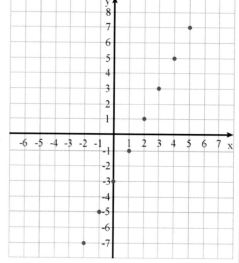

For example, if $x = 0$, then we can calculate the value of y using the equation: $y = 2 \cdot 0 - 3 = -3$. So when $x = 0$ and $y = -3$, the equation is true. The **number pair** $(x, y) = (0, -3)$ is a solution.

Similarly, if x is 3, then $y = 2 \cdot 3 - 3 = 3$. The number pair $(3, 3)$ is *also* a solution.

In this way we could generate an infinite number of solutions. Each solution is a number pair that can be plotted on a coordinate grid.

This table lists some x and y values, plotted at the right, for the equation $y = 2x - 3$:

x	−2	−1	0	1	2	3	4	5
y	−7	−5	−3	−1	1	3	5	7

Notice the pattern in the table and in the graph: as the x-values increase by 1, the y-values increase by 2. The plot shows a pattern, as well: the dots form a line that is rising upwards.

1. Plot the points from the equations for the values of x listed in the table. Graph both (a) and (b) in the same grid.

a. $y = x + 4$

x	−9	−8	−7	−6	−5	−4	−3	−2
y								

x	−1	0	1	2	3	4	5
y							

b. $y = 2x - 1$

x	−3	−2	−1	0	1	2	3	4	5
y									

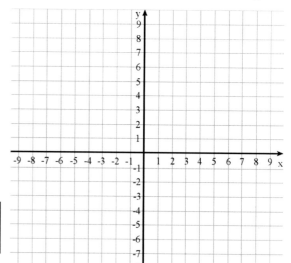

2. Which equation matches the plot on the right?

$y = (\frac{1}{2})x + 1$

$y = (\frac{1}{2})x$

$y = (\frac{1}{2})x - 1$

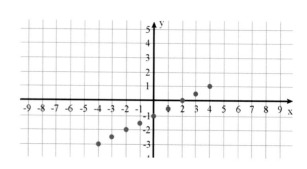

3. Plot the points from the equations for the values of x listed in the table. Graph both (a) and (b) in the same grid.

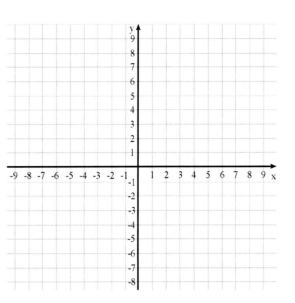

a. $y = -x + 2$

x	−5	−4	−3	−2	−1	0	1	2	3
y									

b. $y = -2x + 1$

x	−4	−3	−2	−1	0	1	2	3	4
y									

Example 1. Is $(5, -2)$ a solution to the equation $y + 3 = 3x$?

Simply substitute the values $x = 5$ and $y = -2$ into the equation, and check if you get a true or false equation (on the right).

Be careful that you substitute the value of x in place of x in the equation, and not in place of y!

No, $(5, -2)$ is not a solution to $y + 3 = 3x$.

$$-2 + 3 \stackrel{?}{=} 3 \cdot 5$$

$$1 \stackrel{?}{=} 15 \quad \times$$

---

4. **a.** Does $(-1, 0)$ fulfill the equation $y = 2x - 1$?

   b. Is $(2, -3)$ a solution to the equation $y - 1 = -x$?

5. Write an equation in the form "$y = mx + b$" (where m and b are constants) that shows how to calculate the value of y from the value of x. Graph the points if it helps you.

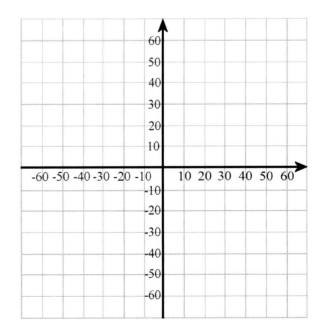

a.

x	−30	−20	−10	0	10	20	30
y	−20	−10	0	10	20	30	40

equation: _____

b.

x	−30	−20	−10	0	10	20	30
y	−60	−40	−20	0	20	40	60

equation: _____

Frequently, we plot an equation for *all* values of *x*, not just for the integer values. This means *x* can be any decimal or fraction, for example 2.56 or 2 3/4. If *x* is a decimal, then *y* probably will be, too.

This would make for a tremendous amount of number pairs to be plotted as dots. We cannot draw that many dots on the grid, so instead we draw a smooth *line* (or a curve for some types of equations). **The line represents all of those specific number pairs (dots).**

Example 2. Graph the equation $y = -2x$.

First, we will plot several points that make the equation $y = -2x$ true, just like we did before.

How do we select those points? We simply choose any *x* values we like and calculate the corresponding *y* values. For example:

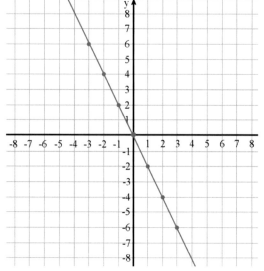

x	−3	−2	−1	0	1	2	3
y	6	4	2	0	−2	−4	−6

We plot those points and see that they fall on a line.

Lastly, we draw a line through the points that extends as far as it can go in both directions. This line now represents all the points that fulfill the equation—even the ones with fractional and decimal coordinates.

Since the plot of this equation is a line, we call the equation a **linear equation**. You can learn more about them in an algebra course.

Graph the equations (as lines). Graph two equations in each coordinate grid.

6. **a.** $y = x + 2$ **c.** $y = (\frac{1}{2})x + 2$

   b. $y = -x + 3$ **d.** $y = 2x - 2$

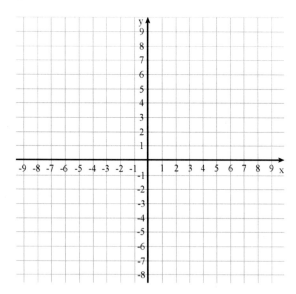

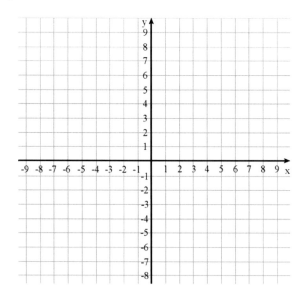

7. How would you check if a given point is on a given line without drawing anything?
   For example, is the point (1, −2) on the line $y = x - 2$?

8. Match each equation with its graph.

$y = (1/2)x - 3$

$y = 4 - x$

$y = -2x + 3$

$y = (2/3)x - 1$

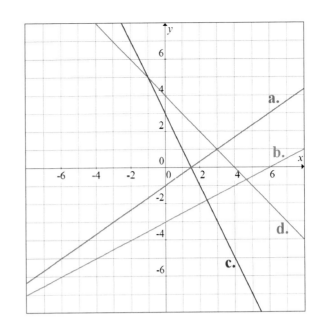

9. Is the point $(-1, 1)$ on the line
   $y = x + 1$?

10. **a.** Plot the equation $y = (1/2)x - 2$.

   b. Plot the equation $y = -3x$.

   c. Plot the equation $y = 6 - x$.

11. Explain two different ways to determine if the
    point $(5, -5)$ is on the line $y = -(1/2)x - 2$.

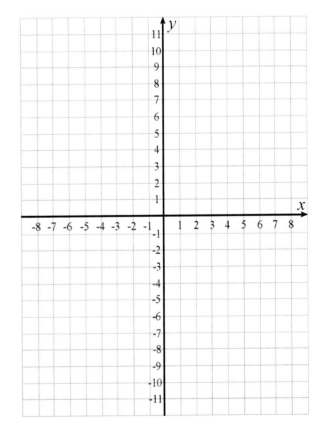

Puzzle Corner

x	−3	−2	−1	0	1	2	3
y	−5	−3	−1	1	3	5	7

Write an equation that relates x and y.

Hint: It is of the form $y = mx + b$ where m and b are integers.

An Introduction to Slope

The **slope** of a line is a number that describes the steepness and direction of its slant or inclination. It is defined as **how many units the *y*-value changes (the "rise") when the *x*-value is changed (the "run") by 1 unit**.

Example 1. What is the slope of the line $y = 2x - 1$?

Looking at the table of some *x* and *y* values, we see that every time the *x*-coordinate increases by 1 unit, the *y*-coordinate increases by 2 units. This means the slope is 2.

x	−4	−3	−2	−1	0	1	2
y	−9	−7	−5	−3	−1	1	3

From any point on the line, draw a one-unit horizontal line segment (the "run") toward the positive *x* direction (right). From the end of that segment, draw another line segment (the "rise") toward the positive *y* direction (up) until you meet the line again. How long is that vertical segment? It is 2 units long. So the slope — the "rise" per unit "run" — is 2.

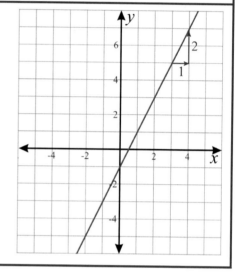

1. Find the slope of the lines. You can use the table, plot the line, or do both.

a. $y = 3x - 2$

x							
y							

b. $y = x + 5$

x							
y							

c. $y = (1/2)x$

x							
y							

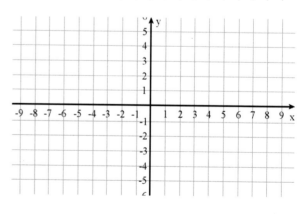

The slope can also be a negative number. In that case, we can say that the line is **decreasing**. Moving from left to right, it goes downwards. Conversely, if the slope is positive, the line is **increasing**.

Example 2. What is the slope of the line $y = -x + 3$?

Every time the x-coordinate increases by 1 unit, the y-coordinate *decreases* by 1 unit. This means the slope is -1.

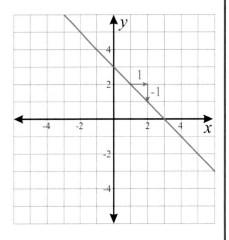

x	−4	−3	−2	−1	0	1	2
y	7	6	5	4	3	2	1

To determine the slope from the graph, again pick any point on the line and draw a one-unit horizontal line segment toward positive x. This time to meet the line you have to draw the second segment downwards. The vertical distance is 1 unit, but the "rise" is negative, toward negative y, so the slope is -1.

2. Find the slope of each line. You can use the table, plot the line, or do both.

a. $y = -3x + 1$

x							
y							

b. $y = -2x$

x							
y							

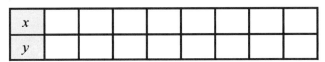

c. $y = 2x - 2$

x							
y							

d. $y = -(1/2)x + 4$

x							
y							

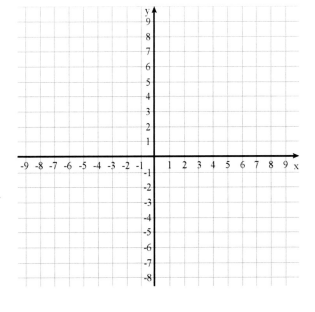

83

Sometimes it is easier to use an increase of some amount other than 1 unit in the x-coordinate. We still use the corresponding change (increase or decrease) in the y-coordinate, but the slope is the ratio of the two changes. So, another way to calculate the slope is:

$$\text{slope} = \frac{\text{change in } y\text{-coordinates}}{\text{change in } x\text{-coordinates}} \quad \text{or} \quad \text{slope} = \frac{\text{rise}}{\text{run}}$$

This is often expressed as "rise over run." The **rise** is the change in the y-coordinates — the change in the vertical direction. The **run** is the change in the x-coordinates — the change in the horizontal direction.

Example 3. Determine the slope from the graph.

As the x-coordinates increase by 3 units (the run), the y-coordinates increase by 5 units (the rise). So the slope is 5/3.

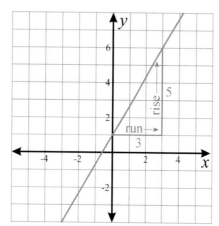

Example 4. Determine the slope from the table.

We can see from the table that each time the x-coordinates increase by 10 units (the run), the y-coordinates *decrease* by 4 units (the rise). Since the rise is negative, so is the slope.

The slope is the ratio of the two changes:
rise / run = −4/10 or −2/5.

x	0	10	20	30	40	50	60
y	16	12	8	4	0	−4	−8

3. Determine the slopes from the graphs. Remember that for a decreasing line, the change in the y-coordinates is negative, which makes the slope negative.

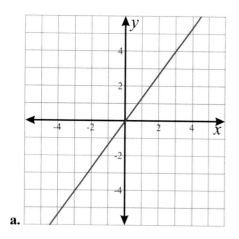

a.

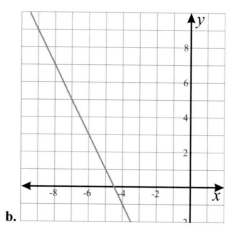

b.

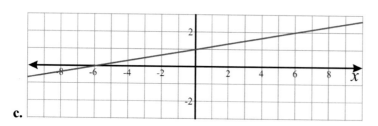

c.

4. Determine the slope of each line from the table or from its graph.

a.

x	−3	−2	−1	0	1	2	3
y	−3 ½	−2	−½	1	2 ½	4	5 ½

b.

x	−4	−2	0	2	4	6	8
y	5	4	3	2	1	0	−1

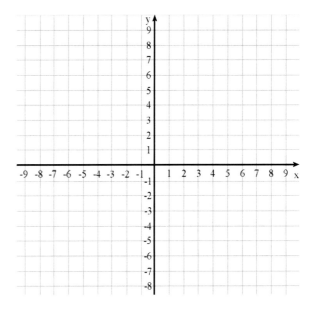

5. Determine the slope of each line. The scaling of the grids is different from that of the grids you have seen in this lesson so far, but the way to find the slope is the same: rise over run.

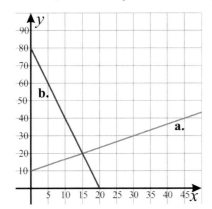

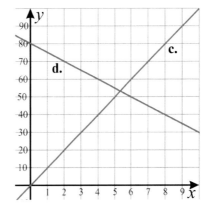

6. Draw two lines with a slope of 3/4. They can be drawn anywhere on the grid; they do not have to go through any specific point.

Check: Your lines should be parallel.

7. Draw two lines with a slope of −3/4.

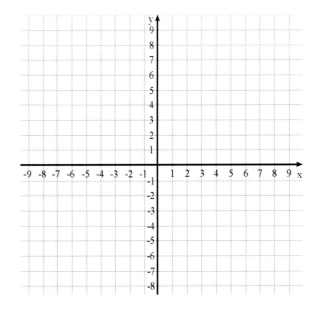

85

8.a. Draw a line that has a slope of 1/2 and
that goes through the point (0, 6).

Hint: Start at the point (0, 6), and draw the
"rise/run diagram" using the ratio 1:2.

b. Draw a line that has a slope of −3 and
that goes through the point (−5, 6).

c. Draw a line that has a slope of 2/3 and
that goes through the point (0, 1).

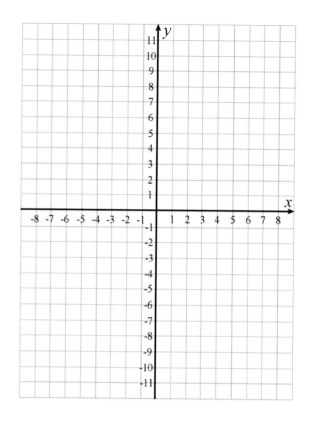

9. Draw any line with a slope of 30.

10. Draw a line that goes through the point (1, 70)
and has a slope of −15.

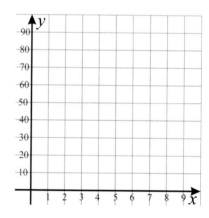

11. **a.** What is the slope of a line that goes
through the points (2, 5) and (3, 8)?

b. What is the slope of a line that goes
through the points (6, 9) and (7, 3)?

12. Sean tried to determine the slope of the line in this graph. He said,
"The slope is 20, because the line goes through the point (10, 20)."
Play teacher and explain what's wrong with Sean's reasoning, how
he can find the correct answer, and what that answer is.

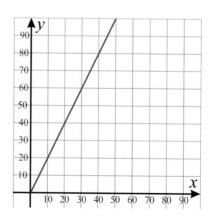

Speed, Time, and Distance

We studied the formula $d = vt$ in the lesson Constant Speed. It tells us how the quantities *distance* (d), *velocity* (v), and *time* (t) are interrelated when an object travels at a constant speed. Their relationship can also be written as $v = d/t$, which you can derive from the common unit for speed, "kilometers per hour."

In this lesson, we explore the relationships between speed, time, and distance in the context of graphing.

Example 1. Harry runs along a 100-meter track at a constant speed. The table below shows his position or distance (d) from the starting line in relation to time (t).

t	0	1	2	3	4	5	6
d	0	5	10	15	20	25	30

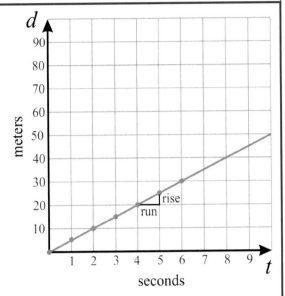

We graph the points and then draw a line through them.

Notice that for each second of time that passes, Harry advances 5 meters. This gives us the "rise/run" relationship that determines the slope of the line.

So, the slope is (5 m)/(1 s), or 5 meters per second. This slope, or change in position over time, is simply Harry's speed.

We can use the slope to relate the quantities t and d in a simple equation: $d = 5t$. Notice that this is simply the formula $d = vt$ with a velocity v of 5 m/s.

In reality, we have to express the velocity in some unit of measure (meters per second in this case), but when we write a formula or an equation, we usually omit the units as a convenience and simply write $d = 5t$ instead of $d = 5$ m/s $\cdot\ t$. However, you still need to include the units in your calculations and final answers.

1. Graph the points. Draw a line through them. Write an equation that relates t and d.

a.

t	0	1	2	3	4	5	6
d	0	4	8	12	16	20	24

equation: _____

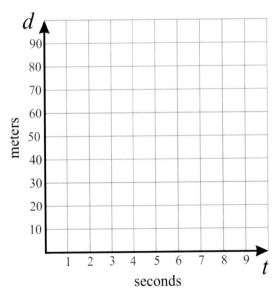

b.

t	0	1	2	3	4	5	6
d	0	7	14	21	28	35	42

equation: _____

c. If the lines represent two runners running with a constant speed, how far from the starting line is each runner when $t = 12$ s?

2. The graph below shows how Henry ran the 100-meter dash.

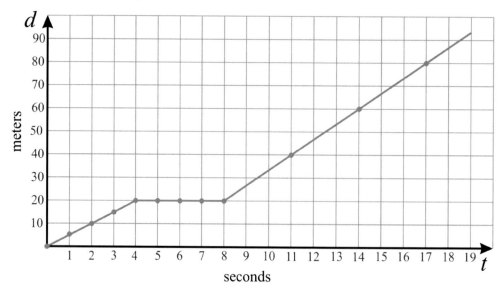

seconds

a. For the first four seconds, Henry runs at a constant speed.
What is his speed?

Also, write an equation relating distance and time in the first four seconds.

b. What happens from the time 4 seconds till 8 seconds?
Look also at the table on the right.
Notice that Henry's position does not change!

t	4	5	6	7	8
d	20	20	20	20	20

c. From 8 seconds and onward, Henry runs at a constant speed again,
but it is different from his earlier speed. What is his speed now?

d. How can we tell *from the graph* that Henry ran at a different speed
at the beginning than at the end?

e. Finish drawing the graph until Henry reaches the 100-meter line.
What is Henry's total running time for the 100-meter race?

3. Sally runs at a constant speed of 5 m/s for six seconds. Then she stops for six seconds to tie her shoelaces. Then she runs to the finish line (at 100 m) at a speed of 7.5 m/s.

a. Plot a graph for the distance Sally runs.

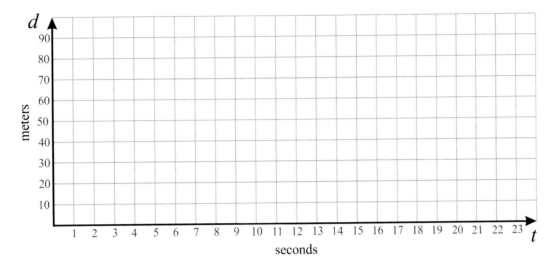

b. What is Sally's total running time for the race?

4. Daisy starts the race, not at the 0-meter mark, but at the 10-meter mark. She runs at a speed of 4 m/s for the first eight seconds. Then she runs the rest of the way (to 100-meter mark) with the speed of 6 m/s.

a. Plot a graph for the distance Daisy runs.

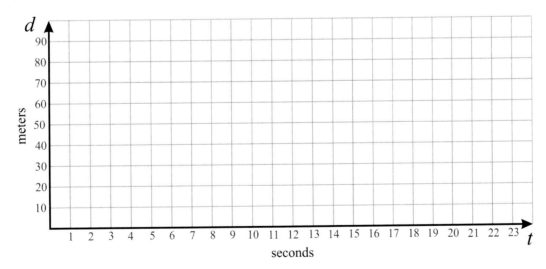

b. How long does she take to run the race?

5. An airplane travels at a constant speed of 800 km/h from Phoenix to Chicago, a distance of 2,320 km.

   a. Write an equation relating the distance (d) it has traveled and the time (t) that has passed.

   b. Plot your equation.

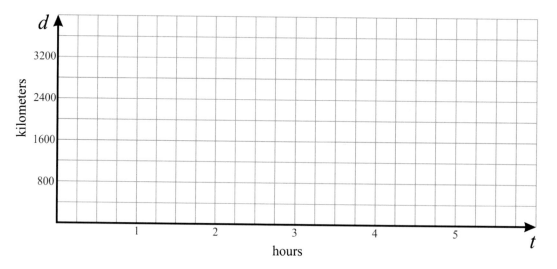

   c. How far will the airplane travel in 2 h 45 minutes?

   d. At what time will the airplane reach Chicago if it left Phoenix at 10:30 am?

6. Tom used a trencher at a steady rate. The graph shows the length of trench he dug in relation to the time he spent.

   a. What is Tom's trenching speed?
   Do not forget to include the units in your answer.

   b. What is the slope of the line in the graph?

   c. Write an equation relating distance (in feet) and time (in minutes).

   d. At the same rate, how many feet of trench could Tom dig in 2 hours?

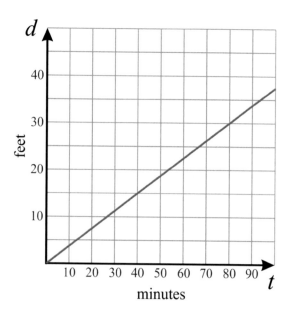

7. Mary drives her tractor at a constant speed. The equation $d = 25t$ tells us the distance, in kilometers, that she travels in t hours.

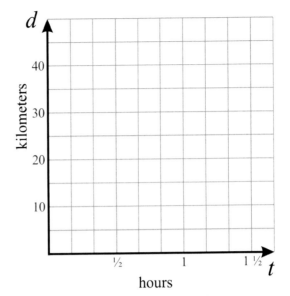

a. What is Mary's speed, in kilometers per hour?
Hint: Find how far Mary can travel in one hour.

b. Plot the equation $d = 25t$.

c. Jerry drove a tractor from his home to a nearby store, a distance of 3 km, in 10 minutes. What was his average speed (in kilometers per hour)?

d. Let's say Jerry drives his tractor at the same constant speed as in (c). Write an equation relating the distance Jerry covers and the time it takes him, and plot it.

e. Use the graph to estimate how much farther Mary can drive than Jerry in 50 minutes.

f. Find the exact answer to (e).

Puzzle Corner

Joe starts the race at the 100-meter mark. He runs *towards the beginning* at a constant speed of 4 m/s until the 30-meter mark. Then he stops there for four seconds. Then he runs at a constant speed of 3 m/s to the starting line. Plot the distance between where Joe started to run and the zero-meter line in relation to time.

Review 2

1. Solve. Check your solutions (as always!).

a. $\quad 1 - 3x \;=\; 17$	**b.** $\qquad 29 \;=\; -6 - 2y$
c. $\dfrac{3x}{8} \;=\; 42$	**d.** $\dfrac{v-2}{7} \;=\; -13$
e. $\dfrac{w}{40} \,-\, 7 \;=\; 19$	**f.** $\dfrac{s+8}{-3} \;=\; -1$

2. Solve each problem in two ways: (1) by writing an equation and (2) by using logical reasoning or a bar model.

a. You bought fifteen bottles of oil for your lawnmower. The shop owner gave you a $14 discount on your entire purchase. The total cost was $130 after the discount. What is the normal price of one bottle of oil?

Equation:

Logical thinking:

b. Three-sevenths of a number is 153. What is the number?

Equation:

Logical thinking:

3. A carpet salesman earns a base salary of $300 a week. He also earns an additional $18 for every carpet he sells.

    a. Write an expression for the salesman's total weekly earnings if he sells n carpets.

    b. How many carpets does the salesman need to sell in order to earn $750 in a week?
       Write an equation and solve it.

4. Solve. Check your solutions.

a. $\quad 2x + 6 + 3x \;=\; 9x - 11$	**b.** $\qquad 2(x + 6) \;=\; 9x - 11$
c. $\qquad 6(5 - w) \;=\; 2(9 - w)$	**d.** $\quad -10(4y + 7) \;=\; -9y$

5. Four adjacent (side-by-side) angles form the line *l*.

   a. Write an equation to solve for the unknown *x*.

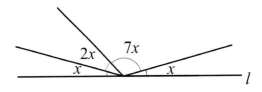

   b. Solve your equation and find the measure of each of the four angles.

6. The total area of a divided room is 200 square feet.
   Find the unknown dimension.

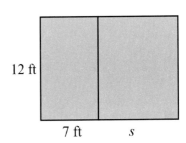

7. Solve the inequalities and plot their solution sets on the number line. You need to write appropriate numbers for the tick marks yourself.

a. $5x - 8 < 22$	**b.** $x + 5 \geq -2$

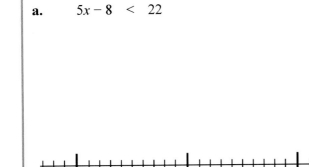

8. Write an equation for this number line diagram and solve it to find the value of the unknown y.

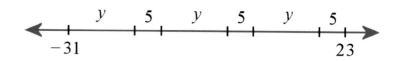

9. An airline has a weight limit of 20 kg for carry-on bags (the luggage passengers carry onto the airplane). Sharon's clothes, personal items, and the carry-on bag itself weigh 9 kg. Besides those, she wants to take a camera that weighs 2.6 kg and as many 0.8-kg bags of nuts as she can. How many bags of nuts can she take?

a. Solve the problem without an equation or inequality.

b. Write an inequality for the problem and solve it.

10. Find the slope of each line. Also, graph the lines.

a. $y = -2x - 1$

x							
y							

Slope: _____

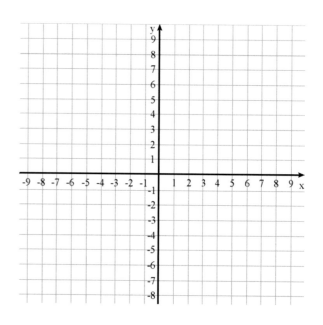

b.

x	−3	−2	−1	0	1	2	3
y	−6 ½	−4	−1 ½	1	3 ½	6	8 ½

Slope: _____

96

11. Draw a line with a slope of 5/6.

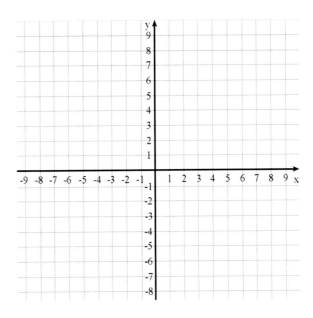

12. Draw a line that has a slope of 3/2 and
that goes through the point (0, 2).

13. An airplane travels at a constant speed of 600 mi/h from New York to Los Angeles, a distance of 2,450 miles.

   a. Write an equation relating the distance (d) it has traveled and the time (t) that has passed.

   b. Plot your equation. Notice that you need to scale the d-axis.

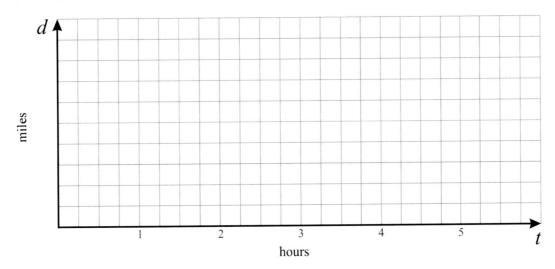

   c. How far will the airplane travel in 1 hour 40 minutes?

Linear Equations Answer Key

Solving Equations, p. 13

1.

a. Equation	Operation to do to both sides
$x + 1 = -4$	-1
$x + 1 + (-1) = -4 + (-1)$	
$x = -5$	

b. Equation	Operation to do to both sides
$x - 1 = -3$	$+1$
$x + (-1) + 1 = -3 + 1$	
$x = -2$	

c. Equation	Operation to do to both sides
$x - 2 = 6$	$+2$
$x + (-2) + 2 = 6 + 2$	
$x = 8$	

d. Equation	Operation to do to both sides
$x + 5 = 2$	-5
$x + 5 + (-5) = 2 + (-5)$	
$x = -3$	

2. a. $x = 10$ b. $x = -1$ c. $x = -7$ d. $x = -9$

3.

a. Equation	Operation to do to both sides
$4x = -12$	$\div 4$
$4x \div 4 = -12 \div 4$	
$x = -3$	

b. Equation	Operation to do to both sides
$(1/3)x = -5$	$\cdot 3$
$(1/3)x \cdot 3 = -5 \cdot 3$	
$x = -15$	

4. a. $p_l = p_e - 8$

   b. $\dfrac{4p}{5} = \$16$

5. $x = -3$

6. Students can solve these equations using guess and check.

a. $x - 7 = 5$
$x = 12$
b. $5 - 8 = x + 1$
$x = -4$
c. $\dfrac{x-1}{2} = 4$
$x = 9$
d. $x^3 = 8$
$x = 2$
e. $-3 = \dfrac{15}{y}$
$y = -5$
f. $5(x + 1) = 10$
$x = 1$

7.

a. Equation	Operation to do to both sides
$3x + 1 = -5$	
$3x + 1 - 1 = -5 - 1$	-1
$3x = -6$	
$3x \div 3 = -6 \div 3$	$\div 3$
$x = -2$	

b. Equation	Operation to do to both sides
$2x - 3 = 4$	
$2x - 3 + 3 = 4 + 3$	$+3$
$2x = 7$	
$2x \div 2 = 7 \div 2$	$\div 2$
$x = 3\ 1/2$	

c. Equation	Operation to do to both sides
$4x + 1 = 13$	
$4x + 1 - 1 = 13 - 1$	-1
$4x = 12$	
$4x \div 4 = 12 \div 4$	$\div 4$
$x = 3$	

Addition and Subtraction Equations, p. 20

1.

a.	$x + 5 = 9$ $-5 -5$ $x = 4$	b.	$x + 5 = -9$ $-5 -5$ $x = -14$
c.	$x - 2 = 3$ $+2 +2$ $x = 5$	d.	$w - 2 = -3$ $+2 +2$ $w = -1$
e.	$z + 5 = 0$ $-5 -5$ $z = -5$	f.	$y - 8 = -7$ $+8 +8$ $y = 1$

2.

a.	$x - 7 = 2 + 8$ $x - 7 = 10$ $+7 +7$ $x = 17$	b.	$x - 10 = -9 + 5$ $x - 10 = -4$ $+10 +10$ $x = 6$
c.	$s + 5 = 3 + (-9)$ $s + 5 = -6$ $-5 -5$ $s = -11$	d.	$t + 6 = -3 - 5$ $t + 6 = -8$ $-6 -6$ $t = -14$

3.

a. $\begin{aligned} -8 &= s + 6 \\ s + 6 &= -8 \\ \underline{-6} \quad &\underline{-6} \\ s &= -14 \end{aligned}$ Check: $-8 \overset{?}{=} -14 + 6$ $-8 = -8$ ✓	b. $\begin{aligned} -2 &= x - 7 \\ x - 7 &= -2 \\ \underline{+7} \quad &\underline{+7} \\ x &= 5 \end{aligned}$ Check: $-2 \overset{?}{=} 5 - 7$ $-2 = -2$ ✓
c. $\begin{aligned} 4 &= s + (-5) \\ s + (-5) &= 4 \\ \underline{+5} \quad &\underline{+5} \\ s &= 9 \end{aligned}$ Check: $4 \overset{?}{=} 9 + (-5)$ $4 = 4$ ✓	d. $\begin{aligned} 2 - 8 &= y + 6 \\ y + 6 &= -6 \\ \underline{-6} \quad &\underline{-6} \\ y &= -12 \end{aligned}$ Check: $2 - 8 \overset{?}{=} -12 + 6$ $-6 = -6$ ✓
e. $\begin{aligned} 5 + x &= -9 \\ x + 5 &= -9 \\ \underline{-5} \quad &\underline{-5} \\ x &= -14 \end{aligned}$ Check: $5 + (-14) \overset{?}{=} -9$ $-9 = -9$ ✓	f. $\begin{aligned} -6 - 5 &= 1 + z \\ z + 1 &= -11 \\ \underline{-1} \quad &\underline{-1} \\ z &= -12 \end{aligned}$ Check: $-6 - 5 \overset{?}{=} 1 + (-12)$ $-11 = -11$ ✓
g. $\begin{aligned} y - (-7) &= 1 - (-5) \\ y + 7 &= 6 \\ \underline{-7} \quad &\underline{-7} \\ y &= -1 \end{aligned}$ Check: $-1 - (-7) \overset{?}{=} 1 - (-5)$ $6 = 6$ ✓	h. $\begin{aligned} 6 + (-2) &= x - 2 \\ x - 2 &= 4 \\ \underline{+2} \quad &\underline{+2} \\ x &= 6 \end{aligned}$ Check: $6 + (-2) \overset{?}{=} 6 - 2$ $4 = 4$ ✓
i. $\begin{aligned} 3 - (-9) &= x + 5 \\ x + 5 &= 12 \\ \underline{-5} \quad &\underline{-5} \\ x &= 7 \end{aligned}$ Check: $3 - (-9) \overset{?}{=} 7 + 5$ $12 = 12$ ✓	j. $\begin{aligned} 2 - 8 &= 2 + w \\ w + 2 &= -6 \\ \underline{-2} \quad &\underline{-2} \\ w &= -8 \end{aligned}$ Check: $2 - 8 \overset{?}{=} 2 + (-8)$ $-6 = -6$ ✓

4.

a. $\begin{aligned} -x &= 6 \\ x &= -6 \end{aligned}$	b. $\begin{aligned} -x &= 5 - 9 \\ -x &= -4 \\ x &= 4 \end{aligned}$
c. $\begin{aligned} 4 + 3 &= -y \\ -y &= 7 \\ y &= -7 \end{aligned}$	d. $\begin{aligned} -2 - 6 &= -z \\ -z &= -8 \\ z &= 8 \end{aligned}$

5. a. $4(s - 1/2) = 12$ and $4(s - 0.5) = 12$

   b. Since $4 \times 3 = 12$, $s - \frac{1}{2} = 3$. So the sides were $3 + \frac{1}{2} = 3\frac{1}{2}$ meters long before they were shortened.

c.
$$
\begin{aligned}
4(s - 1/2) &= 12 \\
s - 1/2 &= 12/4 \\
s - 1/2 &= 3 \\
s - 1/2 + 1/2 &= 3 + 1/2 \\
s &= 3\ 1/2
\end{aligned}
$$

This solution matches this thought process:

Right now, the perimeter is 12 m, so each side is $12 \div 4 = 3$ m. So, before the sides were reduced by 1/2 m, they were 3 1/2 m long.

OR

c.
$$
\begin{aligned}
4(s - 1/2) &= 12 \\
4s - 2 &= 12 \\
4s - 2 + 2 &= 12 + 2 \\
4s &= 14 \\
s &= 3\ 1/2
\end{aligned}
$$

This solution matches this thought process:

If we add the 1/2 meter back in to each side, the perimeter increases by a total of 2 m and becomes 14 meters. So the length of the original side is 14 m perimeter divided by 4, which is 3½ m.

6.

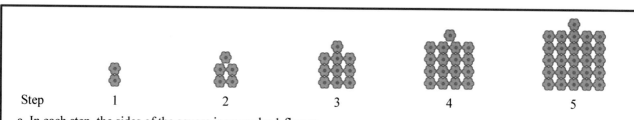

Step   1   2   3   4   5

a. In each step, the sides of the square increase by 1 flower. The one flower on the top does not change.

b. There will be 1,522 flowers in step 39.

c. There will be $n^2 + 1$ flowers in step n.

Addition and Subtraction Equations, cont.

7.

a.	b.
$2 + (-x) = 6 \qquad \mid -2$ $-x = 4$ $\mathbf{x = -4}$ Check: $2 + (-(-4)) \overset{?}{=} 6$ $2 + 4 = 6$ ✓	$8 + (-x) = 7 \qquad \mid -8$ $-x = -1$ $\mathbf{x = 1}$ Check: $8 + (-1) \overset{?}{=} 7$ $7 = 7$ ✓
c.	d.
$-5 + (-x) = 5 \qquad \mid +5$ $-x = 10$ $\mathbf{x = -10}$ Check: $-5 + (-(-10)) \overset{?}{=} 5$ $-5 + 10 = 5$ ✓	$2 + (-x) = -6 \qquad \mid -2$ $-x = -8$ $\mathbf{x = 8}$ Check: $2 + (-8) \overset{?}{=} -6$ $-6 = -6$ ✓
e.	f.
$1 = -5 + (-x) \qquad \mid +5$ $-x = 6$ $\mathbf{x = -6}$ Check: $1 \overset{?}{=} -5 + (-(-6))$ $1 = -5 + 6$ ✓	$2 + (-9) = 8 + (-z) \qquad \mid -8$ $-z = -15$ $\mathbf{z = 15}$ Check: $2 + (-9) \overset{?}{=} 8 + (-15)$ $-7 = -7$ ✓
g.	h.
$-8 + r = -5 + (-7) \qquad \mid +8$ $\mathbf{r = -4}$ Check: $-8 + (-4) \overset{?}{=} -5 + (-7)$ $-12 = 12$ ✓	$2 - (-5) = 2 + 5 + t$ $7 = 7 + t \qquad \mid -7$ $\mathbf{0 = t}$ Check: $2 - (-5) \overset{?}{=} 2 + 5 + 0$ $7 = 7$ ✓

Multiplication and Division Equations, p. 24

1.

a. $\dfrac{8x}{8} = x$	b. $\dfrac{8x}{2} = 4x$	c. $\dfrac{2x}{8} = \dfrac{x}{4}$
d. $\dfrac{-6x}{-6} = x$	e. $\dfrac{-6x}{6} = -x$	f. $\dfrac{6x}{-6} = -x$
g. $\dfrac{6w}{2} = 3w$	h. $\dfrac{6w}{w} = 6$	i. $\dfrac{6w}{-2} = -3w$

Multiplication and Division Equations, cont.

2.

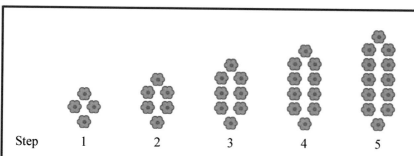

Step   1   2   3   4   5

a. The sides of the shape get one flower taller in each step.
b. There will be 80 flowers in step 39.
c. There will be $2n + 2$ flowers in step n.

3.

a.	b.
$5x = -45$ $\quad\mid \div 5$ $5x/5 = -45/5$ $x = -9$ Check: $5 \cdot (-9) \overset{?}{=} -45$ $-45 = -45$ ✓	$-3y = -21$ $\quad\mid \div (-3)$ $-3y/(-3) = -21/(-3)$ $y = 7$ Check: $-3 \cdot 7 \overset{?}{=} -21$ $-21 = -21$ ✓
c.	d.
$-4 = 4s$ $4s = -4$ $\quad\mid \div 4$ $4s/4 = -4/4$ $s = -1$ Check: $-4 \overset{?}{=} 4 \cdot (-1)$ $-4 = -4$ ✓	$72 = -6y$ $\quad\mid \div (-6)$ $72/(-6) = -6y/(-6)$ $-12 = y$ $y = -12$ Check: $72 \overset{?}{=} -6 \cdot (-12)$ $72 = 72$ ✓

4.

a.	b.
$-5q = -40 - 5$ $-5q = -45$ $\quad\mid \div 5$ $-5q/(-5) = -45/(-5)$ $q = 9$ Check: $-5 \cdot 9 \overset{?}{=} -40 - 5$ $-45 = -45$ ✓	$2 \cdot 36 = -6y$ $72 = -6y$ $\quad\mid \div (-6)$ $72/(-6) = -6y/(-6)$ $y = -12$ Check: $2 \cdot 36 \overset{?}{=} -6 \cdot (-12)$ $72 = 72$ ✓
c.	d.
$3x = -4 + 3 + (-2)$ $3x = -3$ $\quad\mid \div 3$ $3x/3 = -3/3$ $x = -1$ Check: $3 \cdot (-1) \overset{?}{=} -4 + 3 + (-2)$ $-3 = -3$ ✓	$5 \cdot (-4) = -10z$ $-20 = -10z$ $\quad\mid \div (-10)$ $-20/(-10) = \dfrac{-10z}{(-10)}$ $2 = z$ $z = 2$ Check: $5 \cdot (-4) \overset{?}{=} -10 \cdot 2$ $-20 = -20$ ✓

Multiplication and Division Equations, cont.

5.

a. $\dfrac{x}{2} = -45 \qquad \big\| \cdot \mathbf{2}$ $2 \cdot \dfrac{x}{2} = 2 \cdot (-45)$ $x = -90$ Check: $-90/2 \overset{?}{=} -45$ $-45 = -45$ ✓	b. $\dfrac{s}{-7} = -11 \qquad \big\| \cdot (-7)$ $-7 \cdot \dfrac{s}{-7} = -7 \cdot (-11)$ $s = 77$ Check: $77/(-7) \overset{?}{=} -11$ $-11 = -11$ ✓
c. $\dfrac{c}{-7} = 4 \qquad \big\| \cdot (-7)$ $-7 \cdot \dfrac{c}{-7} = -7 \cdot 4$ $c = -28$ Check: $-28/(-7) \overset{?}{=} 4$ $4 = 4$ ✓	d. $\dfrac{a}{-13} = -9 + (-11) \quad \big\| \cdot (-13)$ $-13 \cdot \dfrac{a}{-13} = -13 \cdot (-20)$ $a = 260$ Check: $260/(-13) \overset{?}{=} -9 + (-11)$ $-20 = -20$ ✓

6. a. Let x be the depth at which the shark is swimming. $x = -500/6$; $x = -83\ 1/3$.
   The shark is at a depth of 83 1/3 feet below sea level.

   b. Let c be the total cost. $c/3 = \$21,200$; $c = \$63,600$. The total costs were $63,200.

7.

a. $\dfrac{1}{3}x = -15 \quad \big\| \cdot 3$ $3 \cdot \dfrac{1}{3}x = 3 \cdot (-15)$ $x = -45$ Check: $\dfrac{1}{3}(-45) \overset{?}{=} -15$ $-15 = -15$ ✓	b. $-\dfrac{1}{6}x = -20 \quad \big\| \cdot (-6)$ $(-6) \cdot \left(-\dfrac{1}{6}\right)x = (-6) \cdot (-20)$ $x = 120$ Check: $-\dfrac{1}{6}(120) \overset{?}{=} -20$ $-20 = -20$ ✓	c. $-\dfrac{1}{4}x = 18 \quad \big\| \cdot (-4)$ $-4 \cdot \left(-\dfrac{1}{4}\right)x = -4 \cdot 18$ $x = -72$ Check: $-\dfrac{1}{4}(-72) \overset{?}{=} 18$ $18 = 18$ ✓
d. $-2 = -\dfrac{1}{9}x \quad \big\| \cdot (-9)$ $-9 \cdot (-2) = -9 \cdot \left(-\dfrac{1}{9}\right)x$ $18 = x$ $x = 18$ Check: $-2 \overset{?}{=} -\dfrac{1}{9}(18)$ $-2 = -2$ ✓	e. $-21 = \dfrac{1}{8}x \quad \big\| \cdot 8$ $8 \cdot (-21) = 8 \cdot \dfrac{1}{8}x$ $-168 = x$ $x = -168$ Check: $-21 \overset{?}{=} \dfrac{1}{8} \cdot -168$ $-21 = -21$ ✓	f. $\dfrac{1}{12}x = -7 + 5$ $\dfrac{1}{12}x = -2 \quad \big\| \cdot 12$ $12 \cdot \dfrac{1}{12}x = 12 \cdot (-2)$ $x = -24$ Check: $\dfrac{1}{12}(-24) = -7 + 5$ $-2 = -2$ ✓

105

1. Let s be one side of the square. Equation:

$$4s = 456$$
$$s = 114$$

One side is 114 cm long.

2. Let s be the unknown side of the park. Equation:

$$62s = 4{,}588$$
$$s = 74.$$

The other side is 74 feet long.

3. Let n be the number of boxes John bought. Equation:

$$15n = 165$$
$$n = 11.$$

John bought 11 boxes of screws.

4. Let x be the original length of the candle. Equation:

$$x - 4 \cdot 2 = 6$$
$$x = 14$$

Another possible equation is $2 \cdot 4 + 6 = x$.
The candle was originally 14 cm long.

5. Let w be the weight of the baby dolphin. Equation:

$$(1/12w) = 15$$
$$w = 180$$

The mother dolphin weighs 180 kilograms.

6.

Bar model:	Equation:
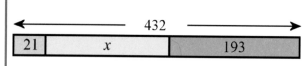 To find the value of x we need to subtract 21 and 193 from 432. $x = 432 - 21 - 193 = 218$	$21 + x + 193 = 432$ $x + 214 = 432 \qquad \big| - 214$ $x = 432 - 214$ $x = 218$

7.

Bar model:	Equation:
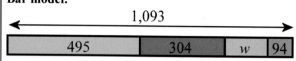 To find the value of w we need to subtract 495, 304, and 94 from 1,093. $w = 1{,}093 - 495 - 304 - 94 = 200$	$495 + 304 + w + 94 = 1{,}093$ $w + 893 = 1{,}093 \qquad \big| - 893$ $w = 1{,}093 - 893$ $w = 200$

1.

a. It will take the caterpillar 1 7/10 hours, which is 1 h 42 min to crawl 34 cm.	b. Father arrives 25 minutes later, which is at 8:05 a.m.
$34 = 20t$ — Flip the sides.	$20 = 48t$ — Flip the sides.
$20t = 34$ — Divide both sides by 20.	$48t = 20$ — Divide both sides by 48
$\dfrac{20t}{20} = \dfrac{34}{20}$ — The factors of 20 in numerator and denominator cancel.	$\dfrac{48t}{48} = \dfrac{20}{48}$ — The factors of 48 in numerator and denominator cancel.
$t = 1\ 7/10$	$t = 5/12$ — In minutes, 5/12 h = 25 min.

2. a. 35/60 = 0.583 hours.   b. 44/60 = 0.733 hours.   c. 2 + 16/60 = 2.267 hours.   d. 4 + 9/60 = 4.154 hours.

3. a. 2 h + 0.4 · 60 min = 2 h 24 min.   b. 0.472 · 60 min ≈ 28 min.
   c. 3 + 3/5 · 60 min = 3 h 36 min.   d. 16/50 · 60 min ≈ 19 min.

4. The bus can travel <u>272 km</u> in 4 hours and 15 minutes.
   Using the equation $d = vt$, we get d = 64 km/h · 4.25 h = 272 km.

5. Sam can run 2.4 miles in approximately <u>14 minutes.</u>
   Using the equation $d = vt$, we get 2.4 mi. = 10 mi/h · t, from which t = 2.4/10 h = 0.24 h = 14.4 minutes.

6. 4 hours 24 minutes.
   For the first half of the distance, we can write the equation, 180 mi. = 90 mi/h · t, from which t = 180/90 h = 2 h.
   For the second half of the distance, we get 180 mi. = 75 mi/h · t, from which t = 180/75 h = 2.4 h.
   In total, the train took 4.4 h = 4 h 24 min.

7. He can jog 2.5 miles.
   From the formula $d = vt$, we get the equation d = 6 mi/h · 25/60 h = 2.5 mi.

8. a. The duck flies 30 miles per hour.
   Since it files 3 miles in 6 minutes, it will fly 30 miles in 60 minutes, which is 1 hour.
   Another way: $v = d/t$ = 3 mi/6 min = 3 mi/(6/60 h) = 3 mi/(1/10 h) = 30 mi/h.

   b. The lion runs 54 kilometers per hour.
   If it runs 0.9 km in 1 minute, then in 60 minutes, it runs 60 · 0.9 km = 54 km.
   Another way: $v = d/t$ = 0.9 km/1 min = 0.9 km/(1/60 h) = 54 km/h.

   c. Henry sleds down the hill 5/6 meters per second.
   $v = d/t$ = 75 m/1.5 min = 75 m/90 s = 15/18 m/s = 5/6 m/s

   d. Rachel swims 3/4 (0.75) kilometers per hour.
   $v = d/t$ = 400 m/32 min = 0.4 km/(32/60 h) = 0.4 km/(8/15 h) = 0.75 km/h.

9. a. His average speed going there was 67.2 kilometers per hour.
   It took him 2 h 14 min = 2 14/60 h ≈ 2.233 h to get there. His average speed was $v = d/t$ = 150 km/2.233 h ≈ 67.2 km/h.

   b. His average speed coming back was 78.2 kilometers per hour.
   It took him 1 h 55 min = 1 55/60 h ≈ 1.917 h to get there. His average speed was $v = d/t$ = 150 km/1.917 h ≈ 78.2 km/h.

   c. His overall average speed was 72.3 kilometers per hour.
   His total travel time was 4 h 9 min = 4 9/60 h = 4.15 h. His average speed was $v = d/t$ = 300 km/4.15 h ≈ 72.3 km/h.

10. It took Jake two hours to drive to his grandparents' place.
   It took Jake two hours and 18 minutes to drive home: t = 150 km/(65 km/h) ≈ 2.308 h ≈ 2 h 18 min.
   So, it took Jake <u>18 minutes longer</u> to drive home than to drive there.

11. We can calculate the average speeds of the two birds in miles per minute and then compare them.
   Seagull: v = 10 mi/24 min = 0.417 mi/min; eagle: 14 mi/30 min = 0.467 mi/min. The eagle is faster.

Constant Speed, cont.

12. a. For the first half of the trip, the train took 80 km/(60 km/h) = 8/6 h = 4/3 h = 1 h 20 min.
For the second half of the trip, the train took 80 km/(240 km/h) = 1/3 h = 20 min.
In total, it took 1 hour 40 minutes for the whole trip.

b. Normally, it would have taken 160 km/(120 km/h) = 16/12 h = 4/3 h = 1 h 20 min for the whole trip.
Notice that doubling the speed for the last half did not make up for the time that was lost traveling the first half at half the normal speed.

c. Its average speed was $v = d/t$ = 160 km/(1 h 40 min) = 160 km/(5/3 h) = 96 km/h.

13. The first 1/3 of the trip: t = 1.5 km/(12 km/h) = 0.125 h = 7.5 min.
The last 2/3 of the trip: t = 3 km/(15 km/h) = 3/15 h = 1/5 h = 12 min.
It will take her a total of <u>19.5 minutes</u>.

14. No, you cannot. Essentially, walking half of the distance at half of your normal speed takes you the exact same time as walking the whole distance at your normal speed, so then it is not possible to make up for the lost time.

Let's say that it is ten miles from your home to the pool.
The first half takes you t = 5 mi/(3 mi/h) = 5/3 h = 1 h 40 min.
The second half takes you t = 5 mi/(12 mi/h) = 5/12 h = 25 min.

Walking with your normal speed, the trip would have taken t = 10 mi/(6 mi/h) = 5/3 mi/h = 1 h 40 min.

15. To make up for the lost time, the airplane should fly at 1,143 km per hour.

In normal conditions, the airplane takes t = 1,600 km/(1,000 km/h) = 1.6 h = 96 min for the whole trip.
In the first 40 minutes = 2/3 h of the trip, the airplane flies 2/3 h · 800 km/h = 533.33 km. This means there is a distance of 1600 km − 533 km = 1,067 km left of the trip.

To make up for the lost time, the airplane has 96 min − 40 min = 56 min = 0.933 h to fly at a faster speed.
Its average speed would need to be 1,067 km/0.933 h = 1,143 km per hour.

Review 1, p. 38

1. a. $x = -13$ b. $x = 4$ c. $x = 10$ d. $z = 9$
e. $x = -132$ f. $q = 120$ g. $c = -1,000$ h. $a = -105$

2. Equation: $3p = 837$. Solution: $p = 279$. One solar panel cost $279.

3. Equation: $s/7 = 187$. Solution: $s = 1,309$. Andrew's salary was $1,309.

4. Substituting the values given in the problem in the formula $d = vt$ gives us the equation 1.2 km = 20 km/h · t.
Solution: t = 1.2 km / (20 km/h) = 0.06 h = 3.6 min = 3 min. 36 sec.

5. $v = d/t$ = 1.5 km/3 min = 1.5 km/(3/60 h) = 1.5 km/(1/20 h) = 30 km/h.

6. The first half: t = 1 mi/12 mph = (1/12) h = 5 minutes. The second half: t = 1 mi/15 mph = (1/15) h = 4 minutes.
It will take Ed 5 + 4 = 9 minutes to get to school.

1.

a. $5x + 2 = 67$ $\quad\mid -2$ $5x = 65$ $\quad\mid \div 5$ $x = 13$ Check: $5 \cdot 13 + 2 \overset{?}{=} 67$ $65 + 2 \overset{?}{=} 67$ $67 = 67$ ✓	b. $3y - 2 = 71$ $\quad\mid +2$ $3y = 73$ $\quad\mid \div 3$ $y = 73/3 = 24\ 1/3$ Check: $3 \cdot (24\ 1/3) - 2 \overset{?}{=} 71$ $73 - 2 \overset{?}{=} 71$ $71 = 71$ ✓
c. $-2x + 11 = 75$ $\quad\mid -11$ $-2x = 64$ $\quad\mid \div (-2)$ $x = -32$ Check: $-2 \cdot (-32) + 11 \overset{?}{=} 75$ $64 + 11 = 75$ ✓	d. $8z - 2 = -98$ $\quad\mid +2$ $8z = -96$ $\quad\mid \div 8$ $x = -12$ Check: $8 \cdot (-12) - 2 \overset{?}{=} -98$ $-96 - 2 = -98$ ✓

2.

a. $\dfrac{x + 6}{5} = 14$ $\quad\mid \cdot 5$ $\dfrac{\cancel{5} \cdot (x + 6)}{\cancel{5}} = 70$ $x + 6 = 70$ $\quad\mid -6$ $x = 64$ Check: $\dfrac{64 + 6}{5} \overset{?}{=} 14$ $\dfrac{70}{5} = 14$ ✓	b. $\dfrac{x + 2}{7} = -1$ $\quad\mid \cdot 7$ $\dfrac{\cancel{7} \cdot (x + 2)}{\cancel{7}} = -7$ $x + 2 = -7$ $\quad\mid -2$ $x = -9$ Check: $\dfrac{-9 + 2}{7} \overset{?}{=} -1$ $\dfrac{-7}{7} = -1$ ✓
c. $\dfrac{x - 4}{12} = -3$ $\quad\mid \cdot 12$ $\dfrac{\cancel{12} \cdot (x - 4)}{\cancel{12}} = -36$ $x - 4 = -36$ $\quad\mid +4$ $x = -32$ Check: $\dfrac{-32 - 4}{12} \overset{?}{=} -3$ $\dfrac{-36}{12} = -3$ ✓	d. $\dfrac{x + 1}{-5} = -21$ $\quad\mid \cdot (-5)$ $\dfrac{\cancel{-5} \cdot (x + 1)}{\cancel{-5}} = 105$ $x + 1 = 105$ $\quad\mid -1$ $x = 104$ Check: $\dfrac{104 + 1}{-5} \overset{?}{=} -21$ $\dfrac{105}{-5} = -21$ ✓

3.

a. $\dfrac{x}{10} + 3 = -2$ $\mid -3$

$\dfrac{x}{10} = -5$ $\mid \cdot 10$

$x = -50$

Check: $\dfrac{-50}{10} + 3 \overset{?}{=} -2$

$-5 + 3 \overset{?}{=} -2$

$-2 = -2$ ✓

b. $\dfrac{x+3}{10} = -2$ $\mid \cdot 10$

$\dfrac{10 \cdot (x+3)}{10} = -20$

$x + 3 = -20$ $\mid -3$

$x = -23$

Check: $\dfrac{-23+3}{10} \overset{?}{=} -2$

$\dfrac{-20}{10} \overset{?}{=} -2$

$-2 = -2$ ✓

4.

a. $\dfrac{x}{7} - 8 = -5$ $\mid +8$

$\dfrac{x}{7} = 3$ $\mid \cdot 7$

$x = 21$

Check: $\dfrac{21}{7} - 8 \overset{?}{=} -5$

$3 - 8 \overset{?}{=} -5$

$-5 = -5$ ✓

b. $\dfrac{x-8}{7} = -5$ $\mid \cdot 7$

$x - 8 = -35$ $\mid +8$

$x = -27$

Check: $\dfrac{-27-8}{7} \overset{?}{=} -5$

$\dfrac{-35}{7} \overset{?}{=} -5$

$-5 = -5$ ✓

5.

a. $1 - 5x = 2$ $\mid -1$

$-5x = 1$ $\mid \div (-5)$

$x = -1/5$

Check: $1 - 5 \cdot (-1/5) \overset{?}{=} 2$

$1 + 1 \overset{?}{=} 2$ ✓

b. $12 - 3y = -6$ $\mid -12$

$-3y = -18$ $\mid \div (-3)$

$y = 6$

Check: $12 - 3 \cdot 6 \overset{?}{=} -6$

$12 - 18 \overset{?}{=} -6$ ✓

5.

c. $\quad 10 = 8 - 4y \qquad \mid -8$ $\quad -4y = 2 \qquad\qquad \mid \div(-4)$ $\quad\quad y = -1/2$ Check: $\quad 10 \stackrel{?}{=} 8 - 4 \cdot (-1/2)$ $\quad\quad 10 \stackrel{?}{=} 8 + 2 \;\checkmark$	d. $\quad 7 = 5 - 3t \qquad \mid -5$ $\quad -3t = 2 \qquad\qquad \mid \div(-3)$ $\quad\quad t = -2/3$ Check: $\quad 7 \stackrel{?}{=} 5 - 3 \cdot (-2/3)$ $\quad\quad 7 \stackrel{?}{=} 5 + 2 \;\checkmark$

6. $2x + 10 = 14$

7. $-9 - 3x = -8.3$

8.

a. $\quad \dfrac{2x}{5} = 8 \qquad\qquad \mid \cdot 5$ $\quad \dfrac{\cancel{5} \cdot 2x}{\cancel{5}} = 40$ $\quad\quad 2x = 40 \qquad\qquad \mid \div 2$ $\quad\quad\; x = 20$ Check: $\quad \dfrac{2 \cdot 20}{5} \stackrel{?}{=} 8$ $\quad\quad \dfrac{40}{5} \stackrel{?}{=} 8$ $\quad\quad\quad 8 = 8 \;\checkmark$	b. $\quad \dfrac{3x}{8} = -9 \qquad\qquad \mid \cdot 8$ $\quad \dfrac{\cancel{8} \cdot 3x}{\cancel{8}} = 8 \cdot (-9)$ $\quad\quad 3x = -72 \qquad\qquad \mid \div 3$ $\quad\quad\; x = -24$ Check: $\quad \dfrac{3 \cdot (-24)}{8} \stackrel{?}{=} -9$ $\quad\quad \dfrac{-72}{8} \stackrel{?}{=} -9$ $\quad\quad\quad -9 \quad -9 \;\checkmark$
c. $\quad 15 = \dfrac{-3x}{10} \qquad \mid \cdot 10$ $\quad 150 = -3x \qquad\qquad \mid \div(-3)$ $\quad -50 = x$ $\quad\quad x = -50$ Check: $\quad 15 \stackrel{?}{=} \dfrac{-3(-50)}{10}$ $\quad\quad 15 \stackrel{?}{=} \dfrac{150}{10}$ $\quad\quad 15 = 15 \;\checkmark$	d. $\quad 2 - 4 = \dfrac{s + 4}{5} \qquad \mid \cdot 5$ $\quad 10 - 20 = s + 4$ $\quad\quad -10 = s + 4 \qquad\qquad \mid -4$ $\quad\quad -14 = s$ $\quad\quad\quad s = -14$ Check: $\quad 2 - 4 \stackrel{?}{=} \dfrac{-14 + 4}{5}$ $\quad\quad -2 = \dfrac{-10}{5} \;\checkmark$

8.

e. $\dfrac{x}{2} - (-16) = -5 \cdot 3$	f. $2 - 4p = -0.5 \quad \big| -2$
$\dfrac{x}{2} + 16 = -15 \qquad \big| -16$	$-4p = -2.5 \quad \big| \div(-4)$
$\dfrac{x}{2} = -31 \qquad \big| \cdot 2$	$p = 0.625$
$x = -62$	
	Check: $2 - 4(0.625) \overset{?}{=} -0.5$
	$2 - 2.5 = -0.5$ ✓
Check: $\dfrac{-62}{2} - (-16) \overset{?}{=} -5 \cdot 3$	
$-31 + 16 = -15$ ✓	

Two-Step Equations: Practice, p. 45

1.

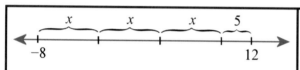	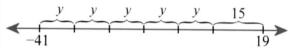
a. $-8 + 3x + 5 = 12$	b. $-41 + 5y + 15 = 19$
$-3 + 3x = 12 \quad \big| +3$	$-26 + 5y = 19 \quad \big| +26$
$3x = 15 \quad \big| \div 3$	$5y = 45 \quad \big| \div 5$
$x = 5$	$y = 9$
Check: $-8 + 3(5) + 5 \overset{?}{=} 12$	Check: $-41 + 5(9) + 15 \overset{?}{=} 19$
$-8 + 15 + 5 \overset{?}{=} 12$	$-41 + 45 + 15 \overset{?}{=} 19$
$12 = 12$ ✓	$19 = 19$ ✓

2.

a. $2 - 5y = -11 \quad \big| -2$	b. $5y - 2 = -11 \quad \big| +2$
$-5y = -13 \quad \big| \div(-5)$	$5y = -9 \quad \big| \div 5$
$y = 13/5 = 2\ 3/5$	$y = -9/5 = -1\ 4/5$
Check: $2 - 5(2\ 3/5) \overset{?}{=} -11$	Check: $5(-1\ 4/5) - 2 \overset{?}{=} -11$
$2 - 13 \overset{?}{=} -11$	$-9 - 2 \overset{?}{=} -11$
$-11 = -11$ ✓	$-11 = -11$ ✓

3.

a.
$$\frac{2x}{7} = -5 \qquad | \cdot 7$$
$$2x = -35 \qquad | \div 2$$
$$x = -17\tfrac{1}{2}$$

Check: $\frac{2(-17\,\frac{1}{2})}{7} \overset{?}{=} -5$

$\frac{-35}{7} \overset{?}{=} -5$

$-5 = -5$ ✓

b.
$$\frac{x+2}{7} = -5 \qquad | \cdot 7$$
$$x + 2 = -35 \qquad | - 2$$
$$x = -37$$

Check: $\frac{-37+2}{7} \overset{?}{=} -5$

$\frac{-35}{7} \overset{?}{=} -5$

$-5 = -5$ ✓

4.

a.
$$20 - 3y = 65 \qquad | - 20$$
$$-3y = 45 \qquad | \div (-3)$$
$$y = -15$$

Check: $20 - 3(-15) \overset{?}{=} 65$

$20 + 45 \overset{?}{=} 65$

$65 = 65$ ✓

b.
$$6z + 5 = -2.2 \qquad | - 5$$
$$6z = -7.2 \qquad | \div 6$$
$$z = -1.2$$

Check: $6(-1.2) + 5 \overset{?}{=} -2.2$

$-7.2 + 5 \overset{?}{=} -2.2$

$-2.2 = -2.2$ ✓

c.
$$\frac{t+6}{-2} = -19 \qquad | \cdot (-2)$$
$$t + 6 = 38 \qquad | - 6$$
$$t = 32$$

Check: $\frac{32+6}{-2} \overset{?}{=} -19$

$\frac{38}{-2} \overset{?}{=} -19$

$-19 = -19$ ✓

d.
$$\frac{y}{6} - 3 = -0.7 \qquad | + 3$$
$$\frac{y}{6} = 2.1 \qquad | \cdot 6$$
$$y = 12.6$$

Check: $\frac{12.6}{6} - 3 \overset{?}{=} -0.7$

$2.1 - 3 \overset{?}{=} -0.7$

$-0.7 = -0.7$ ✓

Two-Step Equations Practice, cont.

5.

a. Equation: Let s be the length of side for each of the three congruent sides. Then: $3s + 1.4\,\text{m} = 7.1\,\text{m}$ $\mid -1.4\,\text{m}$ $3s = 5.7\,\text{m}$ $\mid \div 3$ $s = 1.9\,\text{m}$	Mental math/logical thinking: Subtract the given side length from the total length of the perimeter: 7.1 m − 1.4 m = 5.7 m. Then divide that result by 3 to get the length of any one of the three congruent sides: 5.7 m ÷ 3 = 1.9 m.
b. Equation: Let p be the price of one basket without any discounts. $6p - \$12 = \$46.80$ $\mid + \$12$ $6p = \$58.80$ $\mid \div 6$ $p = \$9.80$	Logical thinking: First add what you paid and the discount: \$46.80 + \$12 = \$58.80. That is the cost of six baskets with the original price. Now, divide that by 6: \$58.80 ÷ 6 = \$9.80 to find the original cost of one basket.
c. Equation: Let x be the unknown number. $\dfrac{2}{5}x = 466$ $\mid \cdot 5$ $2x = 2330$ $\mid \div 2$ $x = 1{,}165$	Logical thinking: Since 466 is two fifths of the original number, divide 466 ÷ 2 = 233 to get one-fifth of the number. Then multiply that by five, and you have the original number: 5 · 233 = 1,165.

Growing Patterns, p. 49

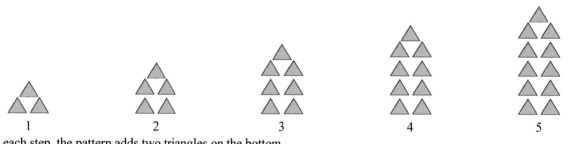

    1   2   3   4   5

1. a. In each step, the pattern adds two triangles on the bottom.
   b. There will be 39 · 2 + 1 = 79 triangles in step 39.
   c. Answers may vary. Check the student's expression. It should be equivalent to $2n + 1$.
      For example, there will be $2n + 1$ or $(2n - 2) + 3$ triangles in step n.
   d. $2n + 1 = 311$
      $2n = 310$
      $n = 155$

    1   2   3   4   5

2. a. In each step, the pattern adds one snowflake on the second and one on the third row.
   b. There will be (2 · 39) + 3 = 81 snowflakes in step 39.
   c. Answers may vary. Check the student's expression. It should be equivalent to $2n + 3$.
      For example, there will be $2n + 2 + 1$ or $2n + 3$ triangles in step n.
   d. $2n + 3 = 301$
      $2n = 298$
      $n = 149$

3.

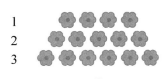

Row	Flowers
1	3 + 1
2	3 + 2
3	3 + 3
4	3 + 4
5	3 + 5
6	3 + 6
7	3 + 7
8	3 + 8
n	3 + n

  a. There will be 3 + n flowers in row n.
  b. There will be 3 + 28 = 31 flowers in the 28th row.
  c. 3 + n = 97
     n = 94
  There will be 97 flowers in row 94.

4. a. There are 2n + 1 flowers in row n.
   b. 2n + 1 = 97
     2n = 96
     n = 48

5. a. The total height of the building is 9n feet.
   b. The total height of the building is 9n + 2 feet.
   c. (1) Using logical thinking: If the total height of the building is 164 ft, then the total
      height of all the stories is 162 ft. There are therefore 162 ÷ 9 = 18 stories.

     (2) Using an equation: 9n + 2 = 164
                  9n = 162
                   n = 18

6. a. 15n + 400

   b. Equation: 15n + 400 = 970
             15n = 570
               n = 38

   He should work 38 hours overtime.

Overtime hours	Total earnings	OR Total earnings
0	400 + 0	400
1	400 + 15	415
2	400 + 30	430
4	400 + 60	460
8	400 + 120	520
10	400 + 150	550
16	400 + 240	640
17	400 + 255	655
30	400 + 450	850
n	400 + 15n	400 + 15n

Puzzle corner:

   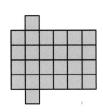 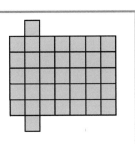

Step   1   2   3   4   5

One way to see this pattern is that it has one individual square at the top and one at the bottom, and in the middle it has a rectangle that is n squares high and n + 2 squares long. So in step n, the pattern has $n(n + 2) + 2$ squares.

In step 59, there will be 59(61) + 2 = **3,601 squares.**

A Variable on Both Sides, p. 53

1.

a.
$$3x + 2 = 2x - 7 \quad | -2x$$
$$3x + 2 - 2x = -7$$
$$x + 2 = -7 \quad | -2$$
$$\mathbf{x = -9}$$

Check: $3 \cdot (-9) + 2 \overset{?}{=} 2 \cdot (-9) - 7$
$$-27 + 2 \overset{?}{=} -18 - 7$$
$$-25 = -25 \checkmark$$

b.
$$9y - 2 = 7y + 5 \quad | -7y$$
$$2y - 2 = 5 \quad | +2$$
$$2y = 7 \quad | \div 2$$
$$\mathbf{y = 7/2}$$

Check: $9 \cdot 7/2 - 2 \overset{?}{=} 7 \cdot 7/2 + 5$
$$31\ 1/2 - 2 \overset{?}{=} 24\ 1/2 + 5$$
$$29\ 1/2 = 29\ 1/2 \checkmark$$

2.

a.
$$11 - 2q = 7 - 5q \quad | +5q$$
$$11 - 2q + 5q = 7$$
$$11 + 3q = 7 \quad | -11$$
$$3q = -4 \quad | \div (-3)$$
$$\mathbf{q = -4/3}$$

Check: $11 - 2 \cdot (-4/3) \overset{?}{=} 7 - 5 \cdot (-4/3)$
$$11 + 2\ 2/3 \overset{?}{=} 7 + 6\ 2/3$$
$$13\ 2/3 = 13\ 2/3 \checkmark$$

b.
$$6z - 5 = 9 - 2z \quad | +2z$$
$$6z - 5 + 2z = 9$$
$$8z - 5 = 9 \quad | +5$$
$$8z = 14 \quad | \div 8$$
$$\mathbf{z = 1\ 3/4}$$

Check: $6 \cdot 1\ 3/4 - 5 \overset{?}{=} 9 - 2 \cdot 1\ 3/4$
$$10\ 1/2 - 5 \overset{?}{=} 9 - 3\ 1/2$$
$$5\ 1/2 = 5\ 1/2 \checkmark$$

c.
$$8x - 12 = -1 - 3x \quad | +3x$$
$$11x - 12 = -1 \quad | +12$$
$$11x = 11 \quad | \div 11$$
$$\mathbf{x = 1}$$

Check: $8 \cdot 1 - 12 \overset{?}{=} -1 - 3 \cdot 1$
$$8 - 12 \overset{?}{=} -1 - 3$$
$$-4 = -4 \checkmark$$

d.
$$-2y - 6 = 20 + 6y \quad | -6y$$
$$-8y - 6 = 20 \quad | +6$$
$$-8y = 26 \quad | \div 8$$
$$\mathbf{y = -3\ 1/4}$$

Check: $-2 \cdot (-3\ 1/4) - 6 \overset{?}{=} 20 + 6 \cdot (-3\ 1/4)$
$$6\ 1/2 - 6 \overset{?}{=} 20 - 19\ 1/2$$
$$1/2 = 1/2 \checkmark$$

e.
$$6w - 6.5 = 2w - 1 \quad | -2w$$
$$4w - 6.5 = -1 \quad | +6.5$$
$$4w = 5.5 \quad | \div 4$$
$$\mathbf{w = 1.375}$$

Check: $6 \cdot 1.375 - 6.5 \overset{?}{=} 2 \cdot 1.375 - 1$
$$8.25 - 6.5 \overset{?}{=} 2.75 - 1$$
$$1.75 = 1.75 \checkmark$$

f.
$$5g - 5 = -20 - 2g \quad | +2g$$
$$7g - 5 = -20 \quad | +5$$
$$\mathbf{7g = -15} \quad | \div 7$$
$$\mathbf{g = -2\ 1/7}$$

Check: $5 \cdot (-2\ 1/7) - 5 \overset{?}{=} -20 - 2 \cdot (-2\ 1/7)$
$$-10\ 5/7 - 5 \overset{?}{=} -20 + 4\ 2/7$$
$$-15\ 5/7 = -15\ 5/7 \checkmark$$

3.

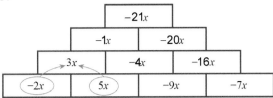

4. Answers may vary. Check the student's work. All of the terms with a variable should come first, in any order, followed by all of the constant terms, in any order.
   a. $2x - 3x - 7 + 11 + 6$
   b. $-s + 15s - 7s - 12 + 9$
   c. $5t - 6t - 2 + (-8)$

5. a. $-x + 10$ b. $7s - 3$ c. $-t - 10$

6. a. $-2x - 7$ b. $-5a$ c. $-5c$ d. $-12x + 1$

7.

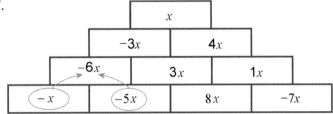

8. a. $8x + 2 + (-3x) + 6 = 5x + 8$ b. $5b - 2 + (-3b) + 9 = 2b + 7$

   c. $-2z + (-3z) + 1 = 1 - 5z$ d. $-4f + 3 + 3f - 4 = -f - 1$

9.

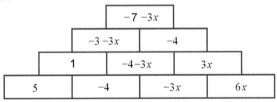

10. a. $1.1y$ b. $-2.6v - 1$ c. $0.1k + 3$

11. This is a challenging equation for this level.

$$
\begin{aligned}
(-1/2)x - 6 + 8x + 7 - x &= 0 \\
(6\ 1/2)x + 1 &= 0 & &| -1 \\
(13/2)x &= -1 & &| \cdot 2 \\
13x &= -2 & &| \div 13 \\
x &= -2/13
\end{aligned}
$$

Check:

$$
\begin{aligned}
(-1/2)(-2/13) - 6 + 8(-2/13) + 7 - (-2/13) &\overset{?}{=} 0 \\
1/13 - 6 - 16/13 + 7 + 2/13 &\overset{?}{=} 0 \\
-13/13 + 1 &\overset{?}{=} 0 \\
0 &= 0 \quad \checkmark
\end{aligned}
$$

12.

a.
$$6x + 3x + 1 = 9x - 2x - 7$$
$$9x + 1 = 7x - 7 \qquad | -7x$$
$$2x + 1 = -7 \qquad | -1$$
$$2x = -8 \qquad | \div 2$$
$$x = -4$$

Check:
$$6(-4) + 3(-4) + 1 \overset{?}{=} 9(-4) - 2(-4) - 7$$
$$-24 - 12 + 1 \overset{?}{=} -36 + 8 - 7$$
$$-35 = -35 \checkmark$$

b.
$$16y - 4y - 3 = -4y - y$$
$$12y - 3 = -5y \qquad | +5y$$
$$17y - 3 = 0 \qquad | +3$$
$$17y = 3 \qquad | \div 17$$
$$y = 3/17$$

Check:
$$16(3/17) - 4(3/17) - 3 \overset{?}{=} -4(3/17) - (3/17)$$
$$48/17 - 12/17 - 3 \overset{?}{=} -12/17 - 3/17$$
$$36/17 - 51/17 \overset{?}{=} -15/17$$
$$-15/17 = -15/17 \checkmark$$

c.
$$-26x + 12x = -18x + 8x - 6$$
$$-14x = -10x - 6 \qquad | +10x$$
$$-4x = -6 \qquad | \div (-4)$$
$$x = 6/4 = 3/2 = 1\ 1/2 \qquad | \div 2$$

Check:
$$-26(3/2) + 12(3/2) \overset{?}{=} -18(3/2) + 8(3/2) - 6$$
$$-39 + 18 \overset{?}{=} -27 + 12 - 6$$
$$-21 = -21 \checkmark$$

d.
$$-9h + 4h + 7 = -2 + 5h + 9h + 8h$$
$$-5h + 7 = -2 + 22h \qquad | -22h$$
$$-27h + 7 = -2 \qquad | -7$$
$$-27h = -9 \qquad | \div (-27)$$
$$h = 1/3$$

Check:
$$-9(1/3) + 4(1/3) + 7 \overset{?}{=} -2 + 5(1/3) + 9(1/3) + 8(1/3)$$
$$-3 + 4/3 + 7 \overset{?}{=} -2 + 5/3 + 3 + 8/3$$
$$5\ 1/3 \overset{?}{=} 1 + 13/3$$
$$5\ 1/3 = 5\ 1/3 \checkmark$$

13.

a.
$$2x - 4 - 7x = -8x + 5 + 2x$$
$$-5x - 4 = -6x + 5 \qquad | +6x$$
$$x - 4 = 5 \qquad | +4$$
$$x = 9$$

Check:
$$2(9) - 4 - 7(9) \overset{?}{=} -8(9) + 5 + 2(9)$$
$$18 - 4 - 63 \overset{?}{=} -72 + 5 + 18$$
$$-49 = -49 \checkmark$$

b.
$$-6 - 4z - 3z = 5z + 8 - z$$
$$-6 - 7z = 4z + 8 \qquad | -4z$$
$$-6 - 11z = 8 \qquad | +6$$
$$-11z = 14 \qquad | \div (-11)$$
$$z = -14/11 = -1\ 3/11$$

Check:
$$-6 - 4(-14/11) \overset{?}{=} 5(-14/11) + 8$$
$$- 3(-14/11) \qquad - (-14/11)$$
$$-6 + 56/11 + 42/11 \overset{?}{=} -70/11 + 8 + 14/11$$
$$-6 + 98/11 \overset{?}{=} -56/11 + 8$$
$$-6 + 8\ 10/11 \overset{?}{=} -5\ 1/11 + 8$$
$$2\ 10/11 = 2\ 10/11 \checkmark$$

13.

<table>
<tr><td>

c.
$$8 - 2m + 5m - 8m = 20 - m + 5m - 2m$$
$$8 - 5m = 20 + 2m \qquad | +5m$$
$$8 = 20 + 7m \qquad | -20$$
$$-12 = 7m$$
$$7m = -12 \qquad | \div 7$$
$$\mathbf{m = -12/7 = -1\ 5/7}$$

Check:

$$8 - 2(-12/7) + 5(-12\ /7) - 8(-12/7) \overset{?}{=} 20 - (-12/7) + 5(-12/7) - 2(-12/7)$$
$$8 + 24/7 - 60/7 + 96/7 \overset{?}{=} 20 + 12/7 - 60/7 + 24/7$$
$$8 + 60/7 \overset{?}{=} 20 - 24/7$$
$$16\ 4/7 = 16\ 4/7 \quad \checkmark$$

</td><td>

d. $\quad -x - x + 2x = 5 - 5x + 9x$
$$0 = 5 + 4x \qquad | -5$$
$$-5 = 4x \qquad | \div 4$$
$$-5/4 = x$$
$$\mathbf{x = -1\ 1/4}$$

Check:

$$-(-5/4) - (-5/4) + 2(-5/4) \overset{?}{=} 5 - 5(-5/4) + 9(-5/4)$$
$$5/4 + 5/4 - 10/4 \overset{?}{=} 5 + 25/4 - 45/4$$
$$0 \overset{?}{=} 5 - 20/4$$
$$0 = 0$$

</td></tr>
<tr><td>

e.
$$-q + 2q - 5q - 6q = 20 - 7 - 9 + q$$
$$-10q = 4 + q \qquad | -q$$
$$-11q = 4 \qquad | \div (-11)$$
$$\mathbf{q = -4/11}$$

Check:

$$-(-4/11) + 2(-4/11) - 5(-4/11) - 6(-4/11) \overset{?}{=} 20 - 7 - 9 + (-4/11)$$
$$4/11 - 8/11 + 20/11 + 24/11 \overset{?}{=} 4 - 4/11$$
$$40/11 \overset{?}{=} 3\ 7/11$$
$$3\ 7/11 = 3\ 7/11 \quad \checkmark$$

</td><td>

f.
$$9 - s + 7 - 9s = 2 - 2s - 11$$
$$16 - 10s = -2s - 9 \qquad | +2s$$
$$16 - 8s = -9 \qquad | -16$$
$$-8s = -25 \qquad | \cdot (-1)$$
$$8s = 25 \qquad | \div 8$$
$$\mathbf{s = 25/8 = 3\ 1/8}$$

Check:

$$9 - (25/8) + 7 - 9(25/8) \overset{?}{=} 2 - 2(25/8) - 11$$
$$16 - 250/8 \overset{?}{=} -9 - 50/8$$
$$16 - 31\ 1/4 \overset{?}{=} -9 - 6\ 1/4$$
$$-15\ 1/4 = -15\ 1/4 \quad \checkmark$$

</td></tr>
</table>

1. Jaime gets 2/5 of the total, which is $820 ÷ 5 · 2 = $\underline{\$328}$. Juan gets the rest, or $820 − $328 = $\underline{\$492}$.

2. a. The two parts, 4 and 7, match the two sides of the rectangle and make up half of the perimeter. So, we have 11 equal parts for half of the perimeter, which is 33 cm. This means one part is 33 cm ÷ 11 = 3 cm. Then, the two sides measure 4 · 3 cm = $\underline{12\ cm}$ and 7 · 3 cm = $\underline{21\ cm}$.

   b. Let the two sides be $4x$ and $7x$. We can write the equation $4x + 7x + 4x + 7x = 66$ or the equation $4x + 7x = 1/2 · 66$, or even $4x + 7x = 33$. Here is the solution of the equation:

$$
\begin{aligned}
4x + 7x &= 1/2 · 66 \\
11x &= 33 \quad\Big|÷ 11 \\
x &= 3
\end{aligned}
$$

   Since the two sides measure $4x$ and $7x$, we substitute $x = 3$ cm in those and get 12 cm and 21 cm.

3. a. First, subtract the shipping charge from the total: $78.20 − $8.75 = $69.45. That is the cost of all five keyboards. Then we get the cost of one by dividing $69.45 ÷ 5 = $13.89.

   b.
$$
\begin{aligned}
5x + \$8.75 &= \$78.20 \quad\Big|-\$8.75 \\
5x &= \$69.45 \quad\Big|÷ 5 \\
x &= \$13.89
\end{aligned}
$$

4. a. First, subtract the shipping cost from the $100: $100 − $7.90 = $92.10. Then divide that by $12 and round the result down to the nearest whole number to get how many mice you can purchase: $92.10 ÷ $12 = 7.675. This means you can purchase 7 mice. You could also estimate: 8 mice would cost 8 · $12 = $96, which would go over $100 with the shipping and handling cost. 7 · $12 = $84, and when the shipping cost is added, the total is still under $100, so you can buy 7 mice.

   b.
$$
\begin{aligned}
\$12x + \$7.90 &= \$100 \quad\Big|-\$7.90 \\
\$12x &= \$92.10 \quad\Big|÷\$12 \\
x &= 7.675
\end{aligned}
$$

   Since you cannot purchase 7.675 mice, this results means you can buy 7 mice.

5. Since the two sides of the rectangle measure x and $x + 2$, and the perimeter is 76 cm, we can write the equation

$$
\begin{aligned}
x + x + 2 + x + x + 2 &= 76 \\
4x + 4 &= 76 \quad\Big|- 4 \\
4x &= 72 \quad\Big|÷ 4 \\
x &= 18
\end{aligned}
$$

So the one side is $x = 18$ cm long, and the other side is $x + 2 = 20$ cm long.

6. a. Since an equilateral triangle has three equal sides, the length of each side is $(24x + 9)/3 = 8x + 3$.

   b. Answers will vary. Check the student's answer. Possible answers listing the four sides include:
   (i) $5x + 1$, $5x + 1$, $5x + 2$, and $5x + 2$; (ii) $8x$, $8x$, $2x + 3$ and $2x + 3$; (iii) 3, 3, $10x$, and $10x$;
   (iv) $6x − 1$, $6x − 1$, $4x + 4$, and $4x + 4$; and so on.

   The answer must meet four conditions:
   (I) Two of the sides must be the same,
   (II) the other two sides must also be the same,
   (III) the four sides must sum to $20x + 6$,
   (IV) each side must have a positive length at least for some values of x.

   Even the sides $−2x$, $−2x$, $12x + 3$, and $12x + 3$ work, because even though x must be a negative number for the side length $−2x$ to be positive, if we choose a value with small absolute value, such as $x = −1/6$, the other two side lengths will still be positive.

Some Problem Solving, cont.

7. a. Since there are 360° in a circle, $5a + 2a + 8a = 360°$.

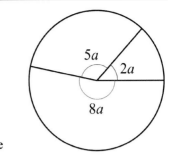

b. We solve the equation from (a):

$$
\begin{aligned}
5a + 2a + 8a &= 360° \\
15a &= 360° \quad \big| \div 15 \\
a &= 24°
\end{aligned}
$$

The three angles measure $5a$, $2a$, and $8a$. Substituting $a = 24$, we get angles that measure 120, 48, and 192 degrees.

8. a. The total angle is half that of a circle, so $2a + 3a + 5a + a = 180°$.

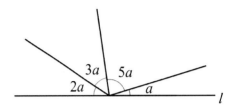

b.
$$
\begin{aligned}
2a + 3a + 5a + a &= 180° \\
11a &= 180° \quad \big| \div 11 \\
a &= 180°/11 \approx 16.36°
\end{aligned}
$$

The angles measure about 32.73°, 49.09°, 81.82, and 16.36°.

Using The Distributive Property, p. 62

1.

<table>
<tr><td>

a.
$$
\begin{aligned}
5(x + 2) &= 85 \\
5x + 10 &= 85 \quad \big| -10 \\
5x &= 75 \quad \big| \div 5 \\
\mathbf{x} &= \mathbf{15}
\end{aligned}
$$

Check:
$$
\begin{aligned}
5(15 + 2) &\overset{?}{=} 85 \\
5(17) &\overset{?}{=} 85 \\
85 &= 85 \checkmark
\end{aligned}
$$

</td><td>

b.
$$
\begin{aligned}
9(y - 2) &= 6y \\
9y - 18 &= 6y \quad \big| -6y \\
3y - 18 &= 0 \quad \big| +18 \\
3y &= 18 \quad \big| \div 3 \\
y &= 6
\end{aligned}
$$

Check:
$$
\begin{aligned}
9(6 - 2) &\overset{?}{=} 6(6) \\
9(4) &\overset{?}{=} 36 \\
36 &= 36 \checkmark
\end{aligned}
$$

</td></tr>
<tr><td>

c.
$$
\begin{aligned}
2(x + 7) &= 6x - 11 \\
2x + 14 &= 6x - 11 \quad \big| -2x \\
14 &= 4x - 11 \quad \big| +11 \\
25 &= 4x \quad \big| \div 4 \\
6\ 1/4 &= x \\
x &= 6\ 1/4
\end{aligned}
$$

Check:
$$
\begin{aligned}
2(6\ 1/4 + 7) &\overset{?}{=} 6(6\ 1/4) - 11 \\
2(13\ 1/4) &\overset{?}{=} 37\ 1/2 - 11 \\
26\ 1/2 &= 26\ 1/2 \checkmark
\end{aligned}
$$

</td><td>

d.
$$
\begin{aligned}
-3z &= 5(z + 9) \\
-3z &= 5z + 45 \quad \big| +3z \\
0 &= 8z + 45 \quad \big| -45 \\
-45 &= 8z \quad \big| \div 8 \\
-45/8 &= z \\
z &= -5\ 5/8
\end{aligned}
$$

Check:
$$
\begin{aligned}
-3(-5\ 5/8) &\overset{?}{=} 5(-5\ 5/8 + 9) \\
16\ 7/8 &\overset{?}{=} 5(3\ 3/8) \\
16\ 7/8 &= 16\ 7/8 \checkmark
\end{aligned}
$$

</td></tr>
</table>

Using the Distributive Property, cont.

1. e.

$$10(x + 9) = -x - 5$$
$$10x + 90 = -x - 5 \qquad | + x$$
$$11x + 90 = -5 \qquad | - 90$$
$$11x = -95 \qquad | \div 11$$
$$x = -95/11 = -8\ 7/11$$

Check:

$$10(-8\ 7/11 + 9) \overset{?}{=} -(-8\ 7/11) - 5$$
$$10(4/11) \overset{?}{=} 8\ 7/11 - 5$$
$$40/11 = 3\ 7/11 \checkmark$$

f.

$$2(5 - s) = 3s - 1$$
$$10 - 2s = 3s - 1 \qquad | + 2s$$
$$10 = 5s - 1 \qquad | + 1$$
$$11 = 5s \qquad | \div 5$$
$$11/5 = s$$
$$s = 2\ 1/5$$

Check:

$$2(5 - 2\ 1/5) \overset{?}{=} 3(2\ 1/5) - 1$$
$$2(2\ 4/5) \overset{?}{=} 6\ 3/5 - 1$$
$$5\ 3/5 = 5\ 3/5 \checkmark$$

2.

a. $6(x - 7) = 72$

One way:	Another way:
$6(x - 7) = 72 \qquad \| \div 6$	$6(x - 7) = 72$
$x - 7 = 12 \qquad \| + 7$	$6x - 42 = 72 \qquad \| + 42$
$x = 19$	$6x = 114 \qquad \| \div 6$
	$x = 19$

b. $10(q - 5) = -60$

One way:	Another way:
$10(q - 5) = -60 \qquad \| \div 10$	$10(q - 5) = -60$
$q - 5 = -6 \qquad \| + 5$	$10q - 50 = -10 \qquad \| + 50$
$q = -1$	$10q = -10 \qquad \| \div 10$
	$q = -1$

3.

a.
$$2(x + 9) = -2x \qquad | \div 2$$
$$x + 9 = -x \qquad | + x$$
$$2x + 9 = 0 \qquad | - 9$$
$$2x = -9 \qquad | \div 2$$
$$x = -4\ 1/2$$

b.
$$\frac{x + 2}{7} = 6x \qquad | \cdot 7$$
$$x + 2 = 42x \qquad | -x$$
$$2 = 41x \qquad \text{Switch the sides.}$$
$$41x = 2 \qquad | \div 41$$
$$x = 2/41$$

122

4.

<table>
<tr><td>

a.
$$2(x + 7) = 3(x - 6)$$
$$2x + 14 = 3x - 18 \quad | -3x$$
$$-x + 14 = -18 \quad | -14$$
$$-x = -32$$
$$\mathbf{x = 32}$$

</td><td>

b.
$$5(y - 2) + 6 = 9$$
$$5y - 10 + 6 = 9$$
$$5y - 4 = 9 \quad | +4$$
$$5y = 13 \quad | \div 5$$
$$\mathbf{y = 2\ 3/5}$$

</td></tr>
<tr><td>

c.
$$2x - 1 = 3(x - 1)$$
$$2x - 1 = 3x - 3 \quad | -3x$$
$$-x - 1 = -3 \quad | +1$$
$$-x = -2$$
$$\mathbf{x = 2}$$

</td><td>

d.
$$3(w - \tfrac{1}{2}) = 6(w + \tfrac{1}{2})$$
$$3w - 1\tfrac{1}{2} = 6w + 3 \quad | -6w$$
$$-3w - 1\tfrac{1}{2} = +3 \quad | +1\tfrac{1}{2}$$
$$-3w = 4\tfrac{1}{2} \quad | \div 3$$
$$\mathbf{w = -1\ \tfrac{1}{2}}$$

</td></tr>
<tr><td>

e.
$$20(x - \tfrac{1}{4}) = x - 2$$
$$20x - 5 = x - 2 \quad | -x$$
$$19x - 5 = -2 \quad | +5$$
$$19x = 3 \quad | \div 19$$
$$\mathbf{x = 3/19}$$

</td><td>

f.
$$8 + 2(7 - v) = 13$$
$$8 + 14 - 2v = 13$$
$$22 - 2v = 13 \quad | -22$$
$$-2v = -9 \quad | \div 2$$
$$\mathbf{v = 4\ \tfrac{1}{2}}$$

</td></tr>
</table>

5. $6(x + 1) = 4x$ or $4x + 12 = 0$.

6.
$$-16 + 3(a + 3) = 29$$
$$-16 + 3a + 9 = 29$$
$$3a - 7 = 29 \quad | +7$$
$$3a = 36 \quad | \div 3$$
$$a = 12$$

7. a. $-5 + 3x = 16$ is modeled as

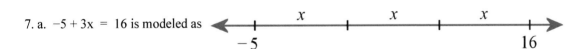

Solution: $-5 + 3x = 16 \quad | +5$
$$3x = 21 \quad | \div 3$$
$$x = 7$$

b. $-5 + 3(x + 2) = 16$ is modeled as

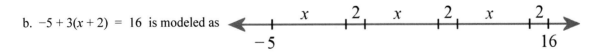

Notice the similarity to the previous problem. The three intervals of x from -5 to 16 in that problem correspond to three intervals of $x + 2$ in this problem. Since x was 7 in the previous problem, it follows that $x + 2 = 7$ in this one, which makes $x = 5$. Or, you can solve the equation the usual way:

$$-5 + 3(x + 2) = 16 \quad | +5$$
$$3(x + 2) = 21 \quad | \div 3$$
$$x + 2 = 7 \quad | -2$$
$$x = 5$$

123

8.

a. $-11(x+2) = -11x - 22$	b. $-3(y+5) = -3y - 15$	c. $-7(x-2) = -7x + 14$
d. $-3(y-20) = -3y + 60$	e. $-10(2x+10) = -20x - 100$	f. $-3(2y+90) = -6y - 270$
g. $-0.5(w+16) = -0.5w - 8$	h. $-0.8(5x-20) = -4x + 16$	i. $-200(0.9x+0.4) = -180x + 80$

9.

a.		b.	
$-2(x+5) = 5x$		$-6(y-2) = 3(y+2)$	
$-2x - 10 = 5x$	$\mid -5x$	$-6y + 12 = 3y + 6$	$\mid -3y$
$-7x - 10 = 0$	$\mid +10$	$-9y + 12 = 6$	$\mid -12$
$-7x = 10$	$\mid \div(-7)$	$-9y = -6$	$\mid \div(-9)$
$x = -1\,3/7$		$y = 2/3$	

c.		d.	
$-0.1(40y+90) = 2$		$3 - t = -5(t - ½)$	
$-4y - 9 = 2$	$\mid +9$	$3 - t = -5t + 2½$	$\mid +5t$
$-4y = 11$	$\mid \div(-4)$	$3 + 4t = 2½$	$\mid -3$
$-y = 2\,3/4$		$4t = -½$	$\mid \div(-6)$
		$t = -1/8$	

10.

a. $-(x+4) = -x - 4$	b. $-(y-9) = -y + 9$	c. $-(x-12) = -x + 12$
d. $-3(a-6) = -3a + 18$	e. $2(-x+5) = -2x + 10$	f. $-4(x-1.2) = -4x + 4.8$
g. $-(4y-20) = -4y + 20$	h. $-(5t+0.9) = -5t - .9$	i. $-(2y+1.4) = -2y - 1.4$
j. $-7(0.2-w) = -1.4 + 7w$	k. $0.1(-60t-7) = -6t - 0.7$	l. $-0.2(9y+4) = -1.8y - 0.8$

11.

Where can you
buy a ruler that
is 3 feet long?   At…

4
A

−10	−7	−5	−7/19
Y	A	R	D

11	1/12	1	−9 ½
S	A	L	E

Since 2 is a solution to each equation, the equation is true when you substitute 2 in place of the variable. Doing that gives you a new equation where the unknown is now the missing number that goes in the ▢. If it makes it easier, you can rewrite the resulting equation with a letter variable in place of the ▢.

a. $(2 - 7) = -10$

      $(-5) = -10$

      $= 2$

b. ▢ $(2 - 5) = 3(2) + 6$

     ▢ $(-3) = 12$

     ▢ $= -4$

c. ▢ $(2 + 8) = 2(2) - 6$

     ▢ $(10) = -2$

      $= -1/5$

d. ▢ $(2) = -6(2 + 1)$

     ▢ $(2) = -18$

     ▢ $= -9$

124

1. a.
$$3 + 2.20n = 25 \quad | -3$$
$$2.20n = 22 \quad | \div 2.20$$
$$n = 10$$

She bought ten ice cream cones.

b.
$$n(3 + 2.83) = 35$$
$$n(5.83) = 35 \quad | \div 5.83$$
$$n = 6.0034$$

He treated six people to coffee and ice cream.

2. Let s be the width of the rectangle. The equation is $28 + s + 28 + s = 144$ or $2s + 56 = 144$.

The solution:
$$2s + 56 = 144 \quad | -56$$
$$2s = 88 \quad | \div 2$$
$$s = 44$$

3. a. $5 \cdot \$12 + \$5$
   b. $n \cdot \$12 + \$5$ which is better written as $12n + 5$.

c.
$$12n + 5 = 173$$
$$12n = 168$$
$$n = 14$$

She can buy 14 bottles with $173.

4. a. $\$12 + \$5 + 4(\$12 + \$1)$ or $\$17 + 4(\$13)$
   b. $\$12 + \$5 + 7(\$12 + \$1)$ or $\$17 + 7(\$13)$
   c. $\$17 + (n - 1)(\$13)$ or written in the standard way, $13(n - 1) + 17$. You can also simplify it to $13n - 13 + 17$ or $13n + 4$.

d.
$$13(n - 1) + 17 = 173$$
$$13n - 13 + 17 = 173$$
$$13n + 4 = 173$$
$$13n = 169$$
$$n = 13$$

She can buy 13 bottles with $173.

5. a. First subtract the cost of the first bottle, $17, from $300. That leaves $283 for the rest of the bottles, each of which cost $13 with the shipping and handling fee. Now divide $283 ÷ $13 ≈ 21.7692. This means she can purchase 21 bottles that cost $13. So in total she can purchase 22 bottles with $300. To check, $1 \cdot \$17 + 21 \cdot \$13 = \$290$. You could also solve this problem by guess and check. You could guess that maybe she can buy 20 bottles, figure out the total cost of that, and then add more bottles until you get near $300 but not go over.

b. Let n be the number of bottles she buys that cost $12 + $1.

$$13n + 17 = 300$$
$$13n = 283$$
$$n \approx 21.7692$$

The solution steps of the equation correspond exactly to the reasoning explained first in (a). However, they do not correspond to the "guess-and-check" method.

Word Problems 2, cont.

6. We can write an equation for the area: length × width = area → 2.5 m · (x + 1.2 m) = 8 m².

$$2.5 \text{ m} \cdot (x + 1.2 \text{ m}) = 8\text{m}^2 \qquad | \div \textbf{2.5 m}$$
$$x + 1.2 \text{ m} = 3.2 \text{ m} \qquad | - \textbf{1.2 m}$$
$$x = 2 \text{ m}$$

So the unknown length is 2 m, and the dimensions of the sandbox are (2 m + 1.2 m) by 2.5 m = 3.2 m by 2.5 m.

Another way to solve this is by using logical reasoning. The area of the smaller rectangle is 2.5 m · 1.2 m = 3 m². That leaves 5 m² for the area of the rectangle on the left. Since one of its sides is 2.5 m, the other must measure 2 m, because 2.5 m · 2 m = 5 m².

7. a. Since three complete trips measure 123.6 km, one complete trip is 123.6 km ÷ 3 = 41.2 km. And, the distance from his home to work is half of that, or 20.6 km. This is 20.6 km − 14 km = 6.6 km longer than his earlier commute.

b. The most difficult part of this problem may be to decide what is the unknown and then write an equation. If we let x be the additional distance beyond 14 km that Tony drives now, then 14 km + x is his total commute one way, 2 · (14 km + x) is the round-trip distance for one commute, and 3 · 2(14 km + x) is the total distance for 3 round trips. So we get the equation:

$$3 \cdot 2(14 + x) = 123.6$$
$$6(14 + x) = 123.6$$
$$14 + x = 20.6$$
$$x = 6.6$$

You could also let the unknown be his commute from home to work. Then the equation would simply be $6y = 123.6$. After solving for y, you would need to subtract the 14 km from it to find how much his commute increased.

8. a. First subtract 360 min − 30 min = 330 min to find the total time spent in classes. Then divide that by 6: 330 min ÷ 6 = 55 min. Each class period is 55 minutes long.

b. Let t be the length of each class period. Then the equation is:

$$6t + 30 = 360 \qquad | \text{subtract the 30-minute study hall}$$
$$6t = 330 \qquad | \text{divide by the 6 class periods}$$
$$t = 55 \qquad | \text{The steps correspond exactly to those in part (a)}$$

Inequalities, p. 71

1. a. ≥ 29 b. ≤ 35 c. > -5 d. < -2

2. a. $a \geq 21$ b. $t \geq 12$ c. $a \geq 55$ d. $b \leq 12$ e. $s \geq 50$

3. Answers will vary. Check the student's answers. Here are some example answers:
   a. She hasn't yet reached 30 years of age.
   b. I've got more than 100 stamps in my collection.
   c. A child has to be at least 8 years old in order to attend the show.
   d. It shouldn't cost more than $60.

4. a. {1, 3, 5, 7}
   b. {6, 8, 10}
   c. {−3, −2, −1, 0}

5.

a.	b.	c.
$3y < 48$ $y < 16$	$y - 8 > 59$ $y > 67$	$2c - 5 \geq 23$ $2c \geq 28$ $c \geq 14$

6.

a. $2x + 12 < 30$
$2x < 18$
$x < 9$

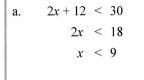

b. $3x - 5 > 83$
$3x > 88$
$x > 29\ 1/3$

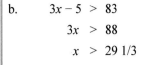

c. $6x - 25 \leq 47$
$6x \leq 72$
$x \leq 12$

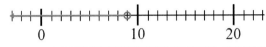

d. $171 \geq 20x - 9$
$180 \leq 20x$
$20x \leq 180$
$x \leq 9$

e. $11a + 5 \leq 2a + 12$
$9a \leq 7$
$a \leq 7/9$

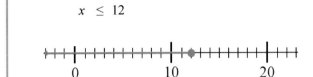

f. $6 + 25y \geq 10y - 9$
$15y \geq -15$
$y \geq -1$

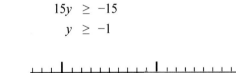

7.

a. $2x - 3 < -9$ $\quad | +3$
$2x < -6$ $\quad | \div 2$
$x < -3$

Check (for example):

$0:\ 2(0) - 3 = -3 > -9$ ✓
$-2:\ 2(-2) - 3 = -4 - 3 = -7 > -9$ ✓
$-4:\ 2(-4) - 3 = -8 - 3 = -11 > -9$ ×
$-10:\ 2(-10) - 3 = -20 - 3 = -23 > -9$ ×

b. $9 + 3x \leq 30$ $\quad | -9$
$3x \leq 21$ $\quad | \div 3$
$x \leq 7$

Check (for example):

$0:\ 9 + 3(0) = 9 \leq 30$ ✓
$6:\ 9 + 3(6) = 9 + 18 = 27 \leq 30$ ✓
$8:\ 9 + 3(8) = 9 + 24 = 31 \leq 30$ ×
$10:\ 9 + 3(10) = 9 + 30 = 39 \leq 30$ ×

7.

c.	d.
$5 + 10x \ < \ 22 \qquad\quad \mid -5$ $10x \ < \ 17 \qquad\quad \mid \div 10$ $x \ < \ 1.7$ Check (for example): 0: $5 + 10(0) = 5 < 22$ ✓ 1: $5 + 10(1) = 5 + 10 = 15 < 22$ ✓ 2: $5 + 10(2) = 5 + 20 = 25 < 22$ × 10: $5 + 10(10) = 5 + 100 = 105 < 22$ ×	$-20 + 3x \ \leq \ 19 \qquad\quad \mid +20$ $3x \ \leq \ 39 \qquad\quad \mid \div 3$ $x \ \leq \ 13$ Check (for example): 0: $-20 + 3(0) = -20 \leq 19$ ✓ 12: $-20 + 3(12) = -20 + 36 = 16 \leq 19$ ✓ 14: $-20 + 3(14) = -20 + 42 = 22 \leq 19$ × 100: $-20 + 3(100) = -20 + 300 = 280 \leq 19$ ×

8.

a. (iii)	b. (iii)
$3x - 7 \ < \ 5 \qquad\quad \mid +7$ $3x \ < \ 12 \qquad\quad \mid \div 3$ $x \ < \ 4$ Notice that the easiest way to solve the problem is to check the endpoint of the graph at $x = 4$, as though the inequalities were written with "≤" instead of "<": (i) $7(4) - 5 = 28 - 5 = 23 \neq 3$ × (ii) $5(4) - 3 = 20 - 3 = 17 \neq 7$ × (iii) $3(4) - 7 = 12 - 7 = 5$ ✓	$4x - 10 \ \geq \ -34 \qquad\quad \mid +10$ $4x \ \geq \ -24 \qquad\quad \mid \div 4$ $x \ \geq \ -6$ Notice that both (i) and (iii) have the correct endpoint at $x = -6$: (i) $4(-6) - 10 = -24 - 10 = -34$ ✓ (ii) $4(-6) + 10 = -24 + 10 = -14 \neq -34$ × (iii) $4(-6) - 10 = -24 - 10 = -34$ ✓ But (i) and (iii) are pointing in opposite directions ("≤" versus "≥"). So the easiest check is to see which one includes 0: (i) $4(0) - 10 = -10 \leq -34$ × (iii) $4(0) - 10 = -10 \geq -34$ ✓

9.

a.	b.
$5x - 17 \ < \ 43$ $5x \ < \ 60$ $x \ < \ 12$ 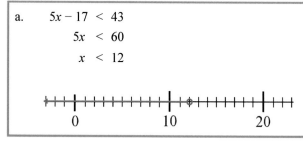	$20x + 18 \ > \ 258$ $20x \ > \ 240$ $x \ > \ 12$ 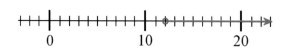

10. a. {1, 2, 3, 4, 5}   b. {12, 14, 16, 18, 20 ...}   c. {13, 14, 15, 16, 17, 18, 19}

11. The second equation, $2y + 6 > -5$, gives a lower bound at $y > -5\,\frac{1}{2}$. The first equation, $15y - 12 < 20$, gives an upper bound at $y < 2\,\frac{2}{15}$. So any three integers from -5 to 2, inclusive, will work: for example, -3, 0, and 2.

Puzzle corner: The number is 6. In the solution below, the only way to get to $x < 5$ is to divide by 6 in the second line. So the missing number must be 6.

$$\blacksquare x - 7 \ < \ 23 \qquad\quad \mid +7$$
$$\blacksquare x \ < \ 30 \qquad\quad \mid \div 6$$
$$x \ < \ 5$$

1. a. $n < 18$. You make at most 17 sales.   b. $n \leq 30$. You make at most 30 sales.

2. Let n be the overtime hours Jeannie works. She earns $\$350 + n \cdot \$18$ in a week, which in mathematics is usually written without the units as $18n + 350$. The inequality is

$$
\begin{aligned}
18n + 350 &\geq 500 && \big| - 350 \\
18n &\geq 150 && \big| \div 18 \\
n &\geq 8\, 1/3
\end{aligned}
$$

Jeannie needs to work 8 1/3 hours or more of overtime to earn **at least $500** in a week.

3. Let n be the number of candle boxes that can be stacked vertically in the shipping container. Including the pad, the total height of the stack is 1 cm $+ n \cdot 6.5$ cm. We can write that expression as $6.5n + 1$ to get the inequality.

$$
\begin{aligned}
6.5n + 1 &\leq 45 && \big| - 1 \\
6.5n &\leq 44 && \big| \div 6.5 \\
n &\leq 6.8
\end{aligned}
$$

With a 1-cm pad on the bottom, she can stack six candle boxes in the shipping container.

4. Let n be the number of pairs of socks you buy. With the $10-dollar coupon, the cost of your order will be $n \cdot \$6.70 - \$10$, which we can write as $6.7n - 10$. The $55 maximum spending limit completes the inequality: $6.7n - 10 \leq 55$.

$$
\begin{aligned}
6.7n - 10 &\leq 55 && \big| + 10 \\
6.7n &\leq 65 && \big| \div 6.7 \\
n &\leq 9.7
\end{aligned}
$$

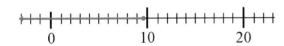

To stay within your $55 limit, the $10 dollar coupon will let you buy at most 9 pairs of socks.

5. Let n be the number of 7.5-hour workdays Joey works. The total number of hours he has toward his apprenticeship is then $21 + n \cdot 7.5$, better written as $7.5n + 21$. The 160-hour minimum gives the inequality $7.5n + 21 \geq 160$.

$$
\begin{aligned}
7.5n + 21 &\geq 160 && \big| - 21 \\
7.5n &\geq 139 && \big| \div 7.5 \\
n &\geq 18.53
\end{aligned}
$$

To finish his apprenticeship Joey will have to work at least 19 workdays.

6. Let n be the number of pages you print. The total cost of printing is then $\$2.45 + n \cdot \$0.03$, or better written as $0.03n + 2.45$. The inequality will then be $0.03n + 2.45 \leq 35$.

$$0.03n + 2.45 \leq 35 \qquad | -2.45$$
$$0.03n \leq 32.55 \qquad | \div 0.03$$
$$n \leq 1,085$$

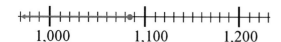

For a total cost of $35 you can print <u>at most 1,085 pages</u>.

Graphing, p. 78

1. a. $y = x + 4$

x	−9	−8	−7	−6	−5	−4	−3	−2
y	−5	−4	−3	−2	−1	0	1	2

x	−1	0	1	2	3	4	5
y	3	4	5	6	7	8	9

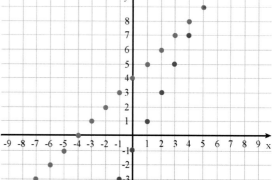

b. $y = 2x - 1$

x	−3	−2	−1	0	1	2	3	4	5
y	−7	−5	−3	−1	1	3	5	7	9

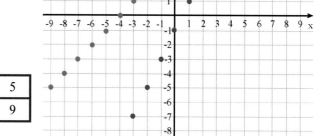

2. $y = (\frac{1}{2})x - 1$. Notice that in this case the substitution $x = 0$ is sufficient to find the answer.

3. a. $y = -x + 2$

x	−5	−4	−3	−2	−1	0	1	2	3
y	7	6	5	4	3	2	1	0	−1

b. $y = -2x + 1$

x	−4	−3	−2	−1	0	1	2	3	4
y	9	7	5	3	1	−1	−3	−5	−7

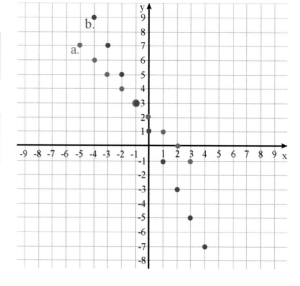

4. a. No   b. No

Graphing, cont.

5. a.

x	−30	−20	−10	0	10	20	30
y	−20	−10	0	10	20	30	40

equation: $y = x + 10$

b.

x	−30	−20	−10	0	10	20	30
y	−60	−40	−20	0	20	40	60

equation: $y = 2x$

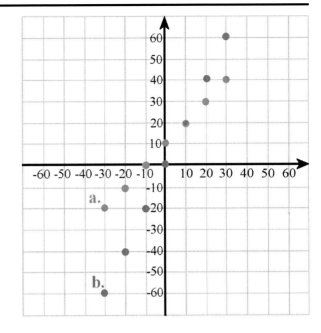

6. a. $y = x + 2$
   b. $y = -x + 3$

   c. $y = (½)x + 2$
   d. $y = 2x - 2$

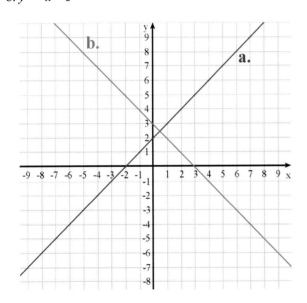

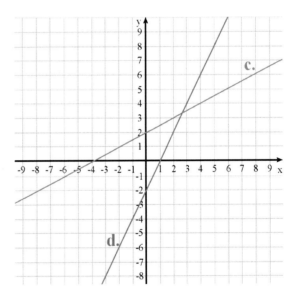

7. Substitute the given values of x and y in the equation to see if the equation is true.
   For example, when we substitute the x and y values of the point (1, −2) into the equation $y = x − 2$, we get $−2 = 1 − 2$. That is a false equation, so the point is <u>not</u> on the line.

8. $y = (1/2)x − 3$ is graph b.
   $y = 4 − x$ is graph d.
   $y = -2x + 3$ is graph c.
   $y = (2/3)x − 1$ is graph a.

   Notice that, since in this case each equation crosses the y-axis (where $x = 0$) in a different place, the simplest way to do the matching is simply to set $x = 0$ in each equation and read off the resulting value of y.

9. No. When we substitute $x = -1$ and $y = 1$ in the equation $y = x + 1$, we get the false equation $1 = -1 + 1$.

131

Graphing, cont.

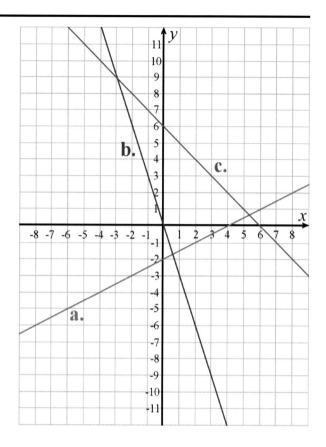

10. a. b. and c. See the graphs on the right.

11. Please check the student's answer.

One way is to draw the line and plot the point in a coordinate grid, and visually check to see if the point falls on the line.

Another way is to substitute the values of the x- and y coordinates of the point into the equation, and check to see if the equation is true. Using the point $x = 5$, $y = -5$, and the equation $y = -(1/2)x - 2$, we get $-5 = -(1/2)5 - 2$, which is not a true equation because it simplifies to $-5 = -4\ 1/2$. So the point is not on the line.

Puzzle corner: $y = 2x + 1$. Either you can guess and check your guess, or you can follow a procedure like this one: Since we're given the form of the equation, $y = ax + b$, and values for x and y, we need to find the values for the constants a and b. If we pick the values for the point where $x = 0$, that makes the equation $y = ax + b$ into $y = 0 + b$, and we'll have our value for b. So the point $(0, 1)$ gives $1 = a(0) + b$, which means that $b = 1$. Now that we know that $b = 1$, substituting in any other point gives us the value for a. Let's pick $(1, 3)$, since both numbers are small and positive. The equation $y = ax + b$ becomes $3 = a(1) + 1$, so $a = 2$. Therefore, the equation is $y = 2x + 1$. We can check our answer by substituting another point, say $(-1, -1)$: $y = 2x + 1$ becomes $-1 = 2(-1) + 1$, which is true, and our equation checks.

An Introduction to Slope, p. 82

1. The x and y values used for the tables will vary. Please check the student's work.

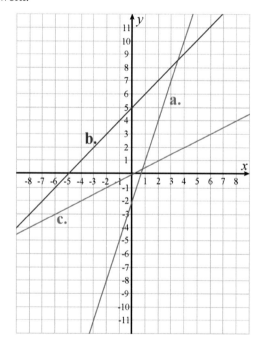

a. The slope is 3.
   $y = 3x - 2$

x	−3	−2	−1	0	1	2	3	4
y	−11	−8	−5	−2	1	4	7	10

b. The slope is 1.
   $y = x + 5$

x	−7	−6	−5	−4	−3	−2	−1	0
y	−2	−1	0	1	2	3	4	5

c. The slope is 1/2.
   $y = (1/2)x$

x	−6	−4	−2	0	2	4	6	8
y	−3	−2	−1	0	1	2	3	4

2. The x and y values used for the tables will vary. Please check the student's work.

a. The slope is −3.

$y = -3x + 1$

x	1	2	3	4	5	6	7	8
y	−2	−5	−8	−11	−14	−17	−20	−23

b. The slope is −2.

$y = -2x$

x	−2	−1	0	1	2	3	4	5
y	4	2	0	−2	−4	−6	−8	−10

c. The slope is 2.

$y = 2x - 2$

x	−6	−5	−4	−3	−2	−1	0	1
y	−14	−12	−10	−8	−6	−4	−2	0

d. The slope is −½.

$y = -(1/2)x + 4$

x	1	2	3	4	5	6	7	8
y	3½	3	2½	2	1½	1	½	0

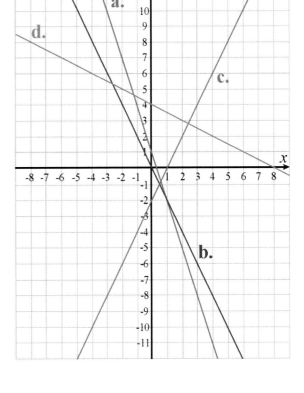

3. a. The slope is 4/3 or 1 1/3.   b. The slope is −2/1 = −2.

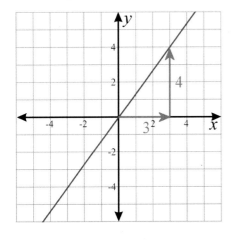   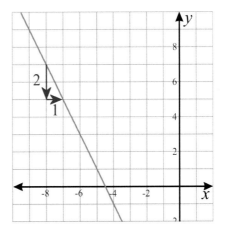

c. The slope is 1/6.

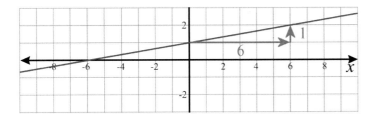

4. a. The slope is 1 ½. Notice the y-values increase by 1 ½ when the x-values increase by 1.
   b. The slope is −1/2. When the y-values *decrease* by 1 x-values increase by 2, so we get a ratio of −1/2.

5. a. The slope is 10/15 = 2/3.
   b. The slope is −20/5 = −4.

c. The slope is 10/1 = 10.
d. The slope is −10/2 = −5.

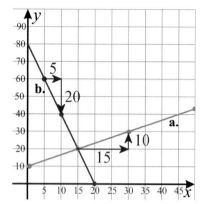 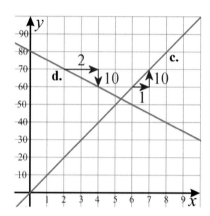

6. Answers will vary. Check the student's work. The student's lines should be parallel to the lines in the graph below:

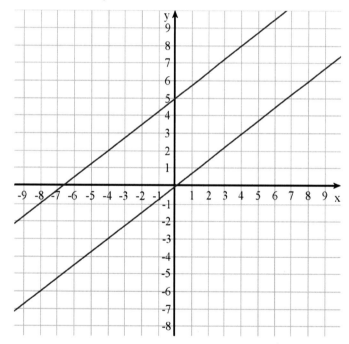

An Introduction to Slope, cont.

7. Answers will vary. Check the student's work. The student's lines should be parallel to the lines in the graph below:

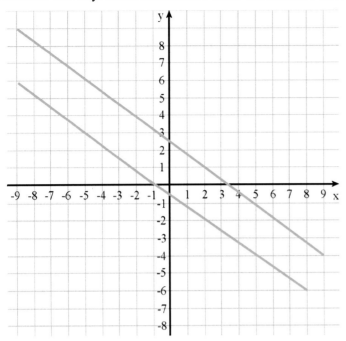

8. a. b. and c.:

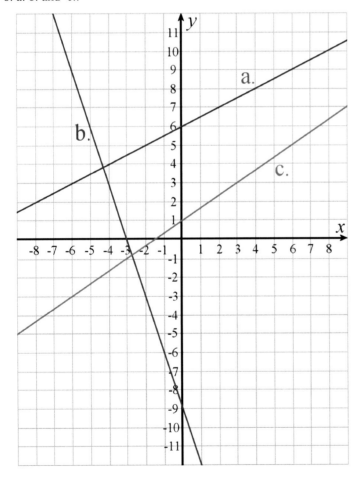

9. Answers will vary. Check the student's work. The student's line should be parallel to the lines in the graph below:

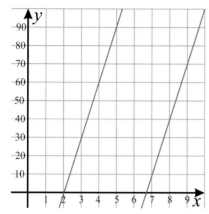

10.

11. a. The slope is 3. The ratio of the change in *y*-coordinates to the change in *x*-coordinates is $(8 - 5)/(3 - 2) = 3/1 = 3$. You can also see that if you plot the points and draw a line through them.
    b. The slope is −6. The ratio of the change in *y*-coordinates to the change in *x*-coordinates is $(3 - 9)/(7 - 6) = -6/1 = -6$. You can also see that if you plot the points and draw a line through them.

12. Sean is calculating the slope based on just one point on the line and possibly simply taking its *y*-coordinate. To determine the slope from the graph, he needs to choose *two* points that are on the line, and then determine the ratio of the change in the *y*-coordinates to the change in the *x*-coordinates. In this case, Sean could use, for example, the points (0, 0) and (10, 20). The change in the *y*-coordinates is $20 - 0 = 20$ and the change in the *x*-coordinates is $10 - 0 = 10$. So, the slope is $20/10 = 2$.

1. a.

t	0	1	2	3	4	5	6
d	0	4	8	12	16	20	24

   equation: $d = 4t$

b.

t	0	1	2	3	4	5	6
d	0	7	14	21	28	35	42

   equation: $d = 7t$

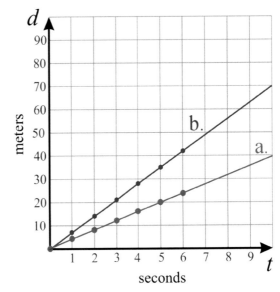

c. Runner *a.* is at a distance of 48 meters when $t = 12$ s.
   Runner *b.* is at a distance of 84 meters when $t = 12$ s.

2. a. For the first four seconds, Henry's speed is 5 meters per second. Equation: $d = 5t$

   b. Maybe Henry is standing still or running in place for four seconds. Maybe he had to stop and tie his shoelace.

   c. From 8 seconds and onward, Henry's speed is 20/3 = 6 2/3 or 6.67 meters per second.

   d. The graph rises at a steeper angle in the end than in the beginning.

   e. His total running time is 20 seconds. You can see that if you continue the graph:

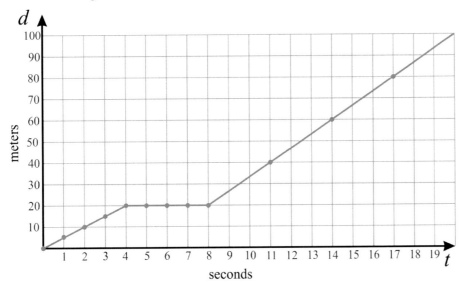

Or you can figure it out. At 8 seconds, Henry is at 20 meters. So he runs 80 more meters to finish the race, and he runs at a speed of 20/3 m/s. The time he takes for that is $t = d/v = 80$ m $/((20/3)$ m/s$) = 3 \cdot 80/20$ s $= 240/20$ s $= 12$ s. So in total he spends 8 s + 12 s = 20 s running the 100-meter race.

3. a.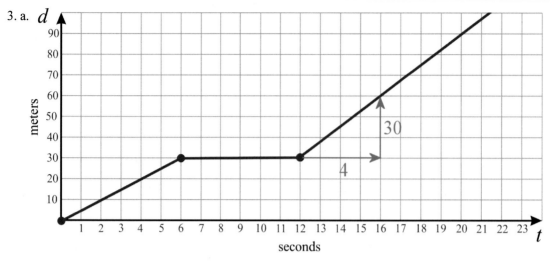

Notice the run/rise diagram in the graph. In the last part of the race, Sally runs at the speed of 7.5 m/s, which is equivalent to running 30 m in 4 s.

b. It took Sally 21 1/3 seconds (about 21.3 s) to run the race.

After the first 12 seconds, she is at the 30-meter mark, so she has 70 meters left after she starts running at the speed of 7.5 m/s. To run those 70 meters takes her 70 m / (7.5 m/s) = 9 1/3 seconds. Her total running time is thus 12 s + 9 1/3 s = 21 1/3 or about 21.3 s.

4. a.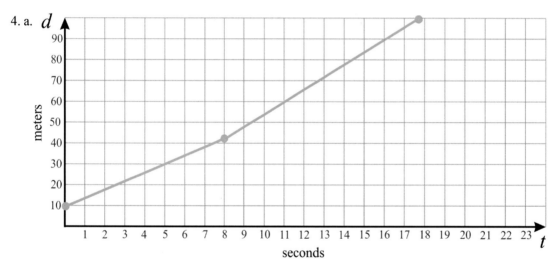

b. It takes Daisy 17 2/3 seconds to run the race.

During the first 8 seconds, she advances to the 42-meter mark (10 m + 8 s · 4 m/s = 42 m). She has 58 meters left when she starts running at the speed of 6 m/s, so that takes her 58/6 = 9 2/3 s. In total her running time is then 8 s + 9 2/3 s = 17 2/3 s.

5. a. $d = 800t$

b.

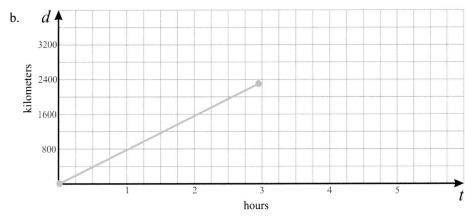

c. The airplane can travel 2,200 kilometers in 2 h 45 minutes. Note that 2 h 45 min is 2.75 hours, so the total distance is 2.75 h · 800 km/h = 2,200 km.

d. The plane will arrive at 1:24 p.m. It takes the airplane 2,320 km /(800 km/h) = 2.9 hours to fly from Phoenix to Chicago. And 0.9 hours is 0.9 · 60 min = 54 min. The total flying time is therefore 2 hours 54 minutes. Lastly add the times: 10:30 a.m. + 2 h 54 min = 13:24 or 1:24 p.m.

6. a. Tom can dig a trench at 3/8 feet per minute or 22.5 ft per hour. From the graph, we can read that he can make 15 ft of trench in 40 minutes, so his trenching rate is 15 ft/40 min. Optionally, if we want to change this into feet per hour, we change 40 minutes to 2/3 hour. The rate in feet per hour is 15 ft/(2/3 h) = 45/2 ft/h = 22 1/2 ft/h.

b. The slope of the line is 15/40 = 3/8. The slope is of course the same as his trenching rate in feet per minute (not the rate in feet per hour because the units in the graph are *feet* and *minutes*).

c. $d = 3/8t$. Again, t is in minutes, and d is in feet.

d. Tom can dig 45 feet of trench in 2 hours. For 2 hours, t is 120 minutes. Using the equation from (c), we get $d = 3/8 \cdot 120 = 45$ feet. You can of course figure this out in other ways as well.

7. a. Mary is driving at 25 km per hour.
   c. Jerry drives at an average speed of 18 km per hour.
   d. $d = 18t$
   e. Estimating from the graph, in 50 minutes Mary can drive about 21 km, and Jerry can drive 15 km, so Mary can drive approximately 6 km farther than Jerry.
   f. In 50 minutes, Mary can drive 25 km/h · (5/6 h) = 20 5/6 or 20.83 km. Jerry can drive 15 km (since he drives 3 km in 10 minutes). So, in 50 minutes, Mary can drive 5 5/6 or 5.83 km farther than Jerry.

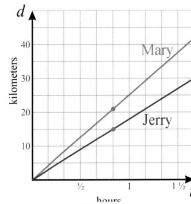

Puzzle corner:

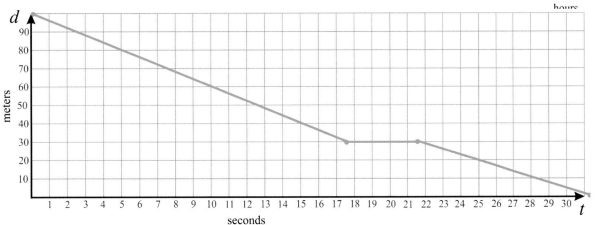

1.

a. $\quad 1 - 3x = 17 \qquad \vert -1$ $\qquad -3x = 16 \qquad \vert \div (-3)$ $\qquad x = -5\ 1/3$ Check: $1 - 3 \cdot (-5\ 1/3) \overset{?}{=} 17$ $1 - (-16) \overset{?}{=} 17$ $1 + 16 \overset{?}{=} 17$ $17 = 17 \qquad \checkmark$	b. $\quad 29 = -6 - 2y \qquad \vert +2y$ $2y + 29 = -6 \qquad \vert -29$ $\qquad 2y = -35 \qquad \vert \div 2$ $\qquad y = -17\ ½$ Check: $29 \overset{?}{=} -6 - 2 \cdot (-17\ ½)$ $29 \overset{?}{=} -6 + 35$ $29 = 29 \qquad \checkmark$
c. $\quad \dfrac{3x}{8} = 42 \qquad \vert \cdot 8$ $\qquad 3x = 336 \qquad \vert \div 3$ $\qquad x = 112$ Check: $\dfrac{3 \cdot 112}{8} \overset{?}{=} 42$ $42 = 42 \qquad \checkmark$	d. $\quad \dfrac{v - 2}{7} = -13 \qquad \vert \cdot 7$ $\qquad v - 2 = -91 \qquad \vert +2$ $\qquad v = -89$ Check: $\dfrac{-89 - 2}{7} \overset{?}{=} -13$ $\dfrac{-91}{7} \overset{?}{=} -13$ $-13 = -13 \quad \checkmark$
e. $\quad \dfrac{w}{40} - 7 = 19 \qquad \vert +7$ $\qquad \dfrac{w}{40} = 26 \qquad \vert \cdot 40$ $\qquad w = 1{,}040$ Check: $\dfrac{1{,}040}{40} - 7 \overset{?}{=} 19$ $26 - 7 \overset{?}{=} 19$ $19 = 19 \qquad \checkmark$	f. $\quad \dfrac{s + 8}{-3} = -1 \qquad \vert \cdot (-3)$ $\qquad s + 8 = 3 \qquad \vert -8$ $\qquad s = -5$ Check: $\dfrac{-5 + 8}{-3} \overset{?}{=} -1$ $\dfrac{3}{-3} \overset{?}{=} -1$ $-1 = -1 \qquad \checkmark$

Review 2, cont.

2.

a.	
Equation:	**Logical thinking:**
Let x be the normal price of one bottle of oil. $15x - 14 = 130$ $15x = 144$ $x = 144/15$ $x = \$9.60$	First, add the total cost and the discount to get the total cost of the oil bottles before the discount: $\$130 + \$14 = \$144$. Then we get the price of one bottle by dividing by 15: $\$144 \div 15 = \$9.60$.
b.	
Equation:	**Logical thinking:**
Let x be the unknown number. $\frac{3}{7}x = 153 \quad \Big| \cdot 7$ $3x = 1071 \quad \Big| \div 3$ $x = 357$	Since three-sevenths of the number is 153, one-seventh of it is $153 \div 3 = 51$. Then, the number itself is seven times that, or $7 \cdot 51 = 357$.

3. a. $300 + 18n$ or $18n + 300$

   b. $\quad 18n + 300 = 750 \quad \Big| -300$

$\qquad\qquad 18n = 450 \quad \Big| \div 18$

$\qquad\qquad\quad n = 25$

Check:

$18 \cdot 25 + 300 \overset{?}{=} 750$

$\quad 450 + 300 = 750 \qquad \checkmark$

He needs to sell 25 carpets in order to earn \$750 in a week.

4.

a.	b.
$2x + 6 + 3x = 9x - 11$ $5x + 6 = 9x - 11 \quad \Big| -6$ $5x = 9x - 17 \quad \Big| -9x$ $-4x = -17 \quad \Big| \div(-4)$ $x = 4\frac{1}{4}$ Check: $2 \cdot (4\frac14) + 6 + 3 \cdot (4\frac14) \overset{?}{=} 9 \cdot (4\frac14) - 11$ $8\frac12 + 6 + 12\frac34 \overset{?}{=} 38\frac14 - 11$ $27\frac14 = 27\frac14 \quad \checkmark$	$2(x+6) = 9x - 11$ $2x + 12 = 9x - 11 \quad \Big| -12$ $2x = 9x - 23 \quad \Big| -9x$ $-7x = -23 \quad \Big| \div(-7)$ $x = 3\,2/7$ Check: $2(3\,2/7 + 6) \overset{?}{=} 9 \cdot (3\,2/7) - 11$ $2(9\,2/7) \overset{?}{=} 29\,4/7 - 11$ $18\,4/7 = 18\,4/7 \quad \checkmark$

4.

c. $6(5-w)$ $=$ $2(9-w)$

$30-6w$ $=$ $18-2w$ $\quad |+2w$

$30-4w$ $=$ 18 $\quad |-30$

$-4w$ $=$ -12 $\quad |\div(-4)$

w $=$ 3

Check:

$6(5-3)$ $\overset{?}{=}$ $2(9-3)$

$6(2)$ $\overset{?}{=}$ $2(6)$

12 $=$ 12 ✓

d. $-10(4y+7)$ $=$ $-9y$

$-40y-70$ $=$ $-9y$ $\quad |+70$

$-40y$ $=$ $-9y+70$ $\quad |+9y$

$-31y$ $=$ 70 $\quad |\div(-31)$

y $=$ $-2\,8/31$

Check:

$-10(4(-2\,8/31)+7)$ $\overset{?}{=}$ $-9(-2\,8/31)$

$-10(-9\,1/31+7)$ $\overset{?}{=}$ $20\,10/31$

$-10(-2\,1/31)$ $\overset{?}{=}$ $20\,10/31$

$20\,10/31$ $=$ $20\,10/31$ ✓

5. a. $x + 2x + 7x + x$ $=$ 180

$11x$ $=$ 180 $\quad |\div 11$

x $=$ $180/11$

x $=$ $16\,4/11 \approx 16.36$

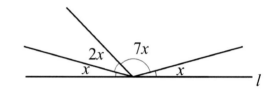

b. The measures of the four angles as fractions are: $x = 16\,4/11°$, $2x = 32\,8/11°$, $7x = 114\,6/11°$, and $x = 16\,4/11°$. However, angles aren't usually measured in elevenths of a degree, so it's more natural to give the answer in decimals. Therefore, the measures of the angles, rounded to two decimal digits, are 16.36°, 32.73°, 114.55°, and 16.36°.

6. One side of the rectangle is 12 ft, and the other side is 7 ft + s. We can write the equation $12(7 + s) = 200$ for the total area.

$12(7+s)$ $=$ 200

$84 + 12s$ $=$ 200 $\quad |-84$

$12s$ $=$ 116 $\quad |\div 12$

s $=$ $9\,2/3$

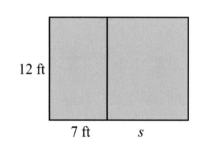

The unknown side measures 9 2/3 ft or 9 ft 8 in.

7.

a. $5x - 8 < 22$

$5x < 30$

$x < 6$

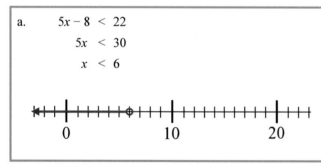

b. $x + 5 \geq -2$

$x \geq -7$

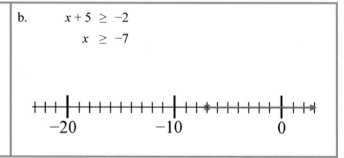

142

8.
$$-31 + 3(y + 5) = 23$$
$$-31 + 3y + 15 = 23$$
$$-16 + 3y = 23 \quad | + 16$$
$$3y = 39 \quad | \div 3$$
$$y = 13$$

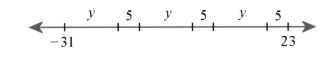

9. a. After the camera, clothes, and personal items she has 20 kg − 2.6 kg − 9 kg = 8.4 kg that she can use for the nuts. Ten bags of nuts weigh 8 kg. Eleven bags weigh 8.8 kg. So she can take only ten bags of nuts.

b.
$$0.8n + 2.6 + 9 = 20$$
$$0.8n + 11.6 = 20 \quad | - 11.6$$
$$0.8n = 8.4 \quad | \div 0.8$$
$$n = 10.5$$

In this situation, we need to round the number of bags down to $n = 10$. (She cannot take 11 bags of nuts; that would put her over the weight limit.

10. a. $y = -2x - 1$

The slope is −2. Data in the table will vary.

x	2	1	0	−1	−2	−3	−4	−5
y	−5	−3	−1	1	3	5	7	9

b. The slope is 2 ½.

x	−3	−2	−1	0	1	2	3
y	−6 ½	−4	−1 ½	1	3 ½	6	8 ½

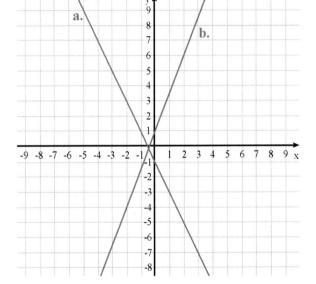

11. Answers will vary. Check the student's answer. The student's line should be parallel to the two purple example lines in the image on the right.

12. See the green line in the image on the right.

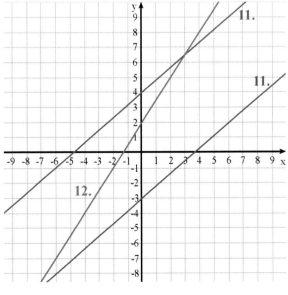

13. a. $d = 600t$. The distance (d) is in miles, and the time (t) is in hours.

b. Notice the line should go through the point (2.5, 1500) on the grid. (At 2 1/2 hours, the plane has flown 1,500 km).

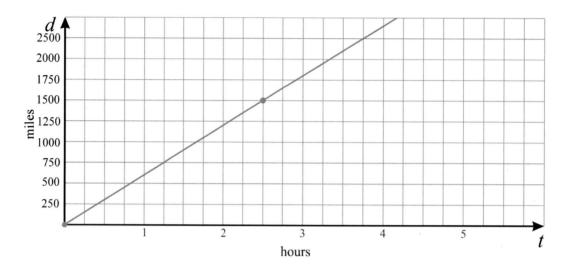

c. Since 1 hour 40 minutes = 1 2/3 hours, in that time the plane has flown $d = (1\ 2/3\ h) \cdot 600\ mi/h = 1,000$ miles.

Linear Equations
Alignment to the Common Core Standards

The table below lists each lesson and next to it the relevant Common Core Standard.

Lesson	Page number	Standards
Solving Equations	13	6.EE.5 6.EE.7 7.EE.4
Addition and Subtraction Equations	20	7.EE.4
Multiplication and Division Equations	24	7.EE.4
Word Problems 1	28	7.EE.4
Constant Speed	31	7.RP.2.c 7.EE.3 7.EE.4
Review 1	38	7.EE.3 7.EE.4
Two-Step Equations	40	7.EE.4
Two-Step Equations: Practice	45	7.EE.4
Growing Patterns	49	7.EE.2 7.EE.4
A Variable on Both Sides	53	8.EE.7
Some Problem Solving	59	7.EE.4
Using the Distributive Property	62	7.EE.4 8.EE.7
Word Problems 2	68	7.EE.3 7.EE.4
Inequalities	71	7.EE.4
Word Problems and Inequalities	76	6.RP.3 7.EE.4
Graphing	78	8.F.3 8.F.4
An Introduction to Slope	82	8.F.2 8.F.3
Speed, Time, and Distance	87	7.RP.2.c 8.EE.5 8.F.5
Review 2	92	7.EE.4 8.EE.7 8.F.3

Made in the USA
Middletown, DE
15 March 2017